Advancing United States–Mexico Binational Sustainability Partnerships

Christopher A. Scott, Jordyn White, and Heather Kreidler, *Editors*

Committee on Sustainability Partnerships in the
U.S.-Mexico Drylands Region

Board on Environmental Change and Society

Division of Behavioral and Social Sciences and Education

A Consensus Study Report of

The National Academies of
SCIENCES • ENGINEERING • MEDICINE

and

Academia Mexicana de Ciencias, Academia de Ingeniería de México, y Academia Nacional de Medicina de México

THE NATIONAL ACADEMIES PRESS
Washington, DC
www.nap.edu

THE NATIONAL ACADEMIES PRESS   500 Fifth Street, NW Washington, DC 20001

This activity was supported by contracts between the National Academies of Sciences, Engineering, and Medicine, the Mexican Academies, the George and Cynthia Mitchell Endowment for Sustainability Sciences at the U.S. National Academy of Sciences (unnumbered), and the National Academy of Sciences W.K. Kellogg Foundation Fund (unnumbered). Support for the work of the Board on Environmental Change and Society is provided primarily by a grant from the National Science Foundation (Award No. BCS-1744000). Any opinions, findings, conclusions, or recommendations expressed in this publication do not necessarily reflect the views of any organization or agency that provided support for the project.

International Standard Book Number-13: 978-0-309-29087-6
International Standard Book Number-10: 0-309-29087-2
Digital Object Identifier: https://doi.org/10.17226/26070

Additional copies of this publication are available from the National Academies Press, 500 Fifth Street, NW, Keck 360, Washington, DC 20001; (800) 624-6242 or (202) 334-3313; http://www.nap.edu.

Spanish language copies of this publication will be available through the Academia Mexicana de Ciencias, km 23.5 Carretera Federal México-Cuernavaca, Calle Cipreses s/n, Col. San Andrés Totoltepec, Tlalpan, 14400 Ciudad de México, México, Tel. +(52 55) 5849-4905, email: aic@unam.mx, http://www.amc.mx.

Copyright 2021 by the National Academy of Sciences. All rights reserved.

Printed in the United States of America

Suggested citation: National Academies of Sciences, Engineering, and Medicine. (2021). *Advancing United States–Mexico Binational Sustainability Partnerships*. Washington, DC: The National Academies Press. https://doi.org/10.17226/26070.

The National Academies of
SCIENCES · ENGINEERING · MEDICINE

The **National Academy of Sciences** was established in 1863 by an Act of Congress, signed by President Lincoln, as a private, nongovernmental institution to advise the nation on issues related to science and technology. Members are elected by their peers for outstanding contributions to research. Dr. Marcia McNutt is president.

The **National Academy of Engineering** was established in 1964 under the charter of the National Academy of Sciences to bring the practices of engineering to advising the nation. Members are elected by their peers for extraordinary contributions to engineering. Dr. John L. Anderson is president.

The **National Academy of Medicine** (formerly the Institute of Medicine) was established in 1970 under the charter of the National Academy of Sciences to advise the nation on medical and health issues. Members are elected by their peers for distinguished contributions to medicine and health. Dr. Victor J. Dzau is president.

The three Academies work together as the **National Academies of Sciences, Engineering, and Medicine** to provide independent, objective analysis and advice to the nation and conduct other activities to solve complex problems and inform public policy decisions. The National Academies also encourage education and research, recognize outstanding contributions to knowledge, and increase public understanding in matters of science, engineering, and medicine.

Learn more about the National Academies of Sciences, Engineering, and Medicine at **www.nationalacademies.org**.

Consensus Study Reports published by the National Academies of Sciences, Engineering, and Medicine document the evidence-based consensus on the study's statement of task by an authoring committee of experts. Reports typically include findings, conclusions, and recommendations based on information gathered by the committee and the committee's deliberations. Each report has been subjected to a rigorous and independent peer-review process and it represents the position of the National Academies on the statement of task.

Proceedings published by the National Academies of Sciences, Engineering, and Medicine chronicle the presentations and discussions at a workshop, symposium, or other event convened by the National Academies. The statements and opinions contained in proceedings are those of the participants and are not endorsed by other participants, the planning committee, or the National Academies.

For information about other products and activities of the National Academies, please visit www.nationalacademies.org/about/whatwedo.

 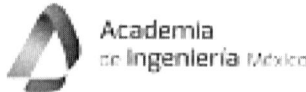

The **Academia Mexicana de Ciencias** (Mexican Academy of Sciences) was established in 1959 as a nonprofit, nongovernmental organization to promote scientific culture in society and to advise the nation on issues related to science and technology. Members are elected by their peers for their distinguished contributions to research. Dr. Susana Lizano-Soberón is president. Website: www.amc.mx

The **Academia de Ingeniería de México** (Mexican Academy of Engineering) created in 1972, is a nonprofit organization that brings together experts with a great sense of social responsibility, who have excelled in practice, research, and teaching of engineering, and contribute to the sustainable development of Mexico. Members are elected by their peers for their contributions to engineering. Dr. Agustín Álvarez-Icaza Longoría is president. Website: www.ai.org.mx

The **Academia Nacional de Medicina de México** (National Academy of Medicine of Mexico), established in 1864, is a nonprofit organization that promotes teaching and research in the field of medicine, and gives advice to professionals, health authorities, and the general public. Members are elected by their peers for their contributions to research and teaching in medicine and public health. Dr. José Halabe Cherem is president. Website: www.anmm.org.mx

COMMITTEE ON SUSTAINABILITY PARTNERSHIPS IN THE U.S.–MEXICO DRYLANDS REGION

CHRISTOPHER A. SCOTT (*Chair*), Udall Center for Studies in Public Policy, University of Arizona
BERNARD AMADEI, University of Colorado Boulder (resigned on 3/20/2020)
ANTHONY BEBBINGTON, Clark University Graduate School of Geography
ROBERT BULLARD, Texas Southern University (resigned on 4/6/2020)
ALFONSO ANDRÉS CORTEZ-LARA, El Colegio de la Frontera Norte-Sede Mexicali
ALMA COTA DE YÁÑEZ, FESAC Fundación del Empresariado, Sonorense, A.C.
HALLIE EAKIN, School of Sustainability, Arizona State University
CONSTANTINO DE JESÚS MACÍAS GARCIA, Instituto de Ecología, Universidad Nacional Autónoma de México
NATALIA MARTÍNEZ-TAGÜEÑA, Instituto Potosino de Investigación Científica y Tecnológica
BENJAMIN L. PRESTON, RAND Corporation (resigned on 9/28/2020)
ROGER S. PULWARTY, National Oceanic and Atmospheric Administration
EXEQUIEL ROLÓN, Fresnillo PLC
KELLY TWOMEY SANDERS, Civil and Environmental Engineering, University of Southern California

Consultant

ELISABETH HUBER-SANNWALD, Instituto Potosino de Investigación Científica y Tecnológica

Staff

JORDYN WHITE, *Study Director*
TOBY WARDEN, *Board Director*
ADAM K. JONES, *Senior Program Assistant*
TINA M. LATIMER, *Program Coordinator*
HEATHER KREIDLER, *Project Consultant*
JOSÉ FRANCO, *Project Consultant* (*Mexican Academies Representative*)
JOSÉ LUIS MORÁN, *Project Consultant* (*Mexican Academies Representative*)
RENATA VILLALBA, *Program Associate* (*Mexican Academies Representative*)

BOARD ON ENVIRONMENTAL CHANGE AND SOCIETY

KRISTIE L. EBI (*Chair*), Center for Health and the Global Environment (CHanGE), University of Washington, Seattle
HALLIE C. EAKIN, School of Sustainability, Arizona State University
LORI M. HUNTER, Institute of Behavioral Science, University of Colorado, Boulder
KATHARINE L. JACOBS, Center for Climate Adaptation Science and Solutions and Department of Soil, Water, and Environmental Science, University of Arizona
MICHAEL ANTHONY MENDEZ, Department of Urban Planning and Public Policy, University of California, Irvine
RICHARD G. NEWELL, Resources for the Future, Washington, DC
ASEEM PRAKASH, College of Arts and Sciences, University of Washington, Seattle
MAXINE L. SAVITZ, Technology/Partnership Honeywell Inc. (retired), Los Angeles, CA
MICHAEL P. VANDENBERGH, School of Law, Vanderbilt University
JALONNE L. WHITE-NEWSOME, Empowering a Green Environment and Economy, LLC, Troy, MI
CATHY L. WHITLOCK, Paleoecology Lab, Montana State University
ROBYN S. WILSON, School of Environment and Natural Resources, The Ohio State University

TOBY WARDEN, *Director*

Reviewers

This Consensus Study Report was reviewed in draft form by individuals chosen for their diverse perspectives and technical expertise. The purpose of this independent review is to provide candid and critical comments that will assist the National Academies of Sciences, Engineering, and Medicine in making each published report as sound as possible and to ensure that it meets the institutional standards for quality, objectivity, evidence, and responsiveness to the study charge. The review comments and draft manuscript remain confidential to protect the integrity of the deliberative process.

We thank the following individuals for their review of this report: Elena Centeno García, UNAM Center for Mexican Studies, University of Arizona, Centro de Estudios Mexicanos UNAM Tucson, Universidad Nacional Autónoma de México; Katharine L. Jacobs, Center for Climate Adaptation Science and Solutions, Arizona Institutes for Resilience, University of Arizona; Stephen P. Mumme, Department of Political Science, Colorado State University; Nicolás Pineda-Pablos, Department of Government and Public Policy, El Colegio de Sonora, México; Alexis Racelis, Department of Biology, College of Sciences, The University of Texas Rio Grande Valley; and Sandi Rosenbloom, School of Architecture, The University of Texas at Austin.

Although the reviewers listed above provided many constructive comments and suggestions, they were not asked to endorse the conclusions or recommendations of this report nor did they see the final draft before its release. The review of this report was overseen by Susan Hanson, Department of Geography, Clark University, and Arun Agrawal, School for Environment and Sustainability, University of Michigan. They were responsible for making certain that an independent examination of this report was carried out in accordance with the standards of the National Academies and that all review comments were carefully considered. Responsibility for the final content rests entirely with the authoring committee and the National Academies.

Preface

Individually, each country's National Academies bring together expertise in science, engineering, and medicine to devise evidence-based solutions to pressing national challenges. Less frequently, challenges that are global or transboundary in nature may be addressed through national efforts. And in even rarer circumstances, challenges that are identified as being binational in their ambit and that require binational expertise for their solutions are the subject of collaboration by two countries' National Academies. This consensus study is one such rare case.

The present report and the process behind it represent a pioneering example of binational cooperation, in which the U.S. National Academies of Sciences, Engineering, and Medicine and the Mexican Academy of Sciences, Academy of Engineering, and National Academy of Medicine jointly identified drylands sustainability as a challenge that affects an extensive region including but not limited to the two nations' border. More importantly, both countries' National Academies recognized that diagnosis, assessment, engagement, and solution needed to be not just binational but also interdisciplinary, involving experts with varied training, as well as transdisciplinary, building on expertise from civil society and the private sector. Further, to demonstrate its global relevance, the study assesses U.S. and Mexican challenges in the context of global sustainable development as defined by the United Nations' Sustainable Development Goals, and in particular Goal 17, which calls for multi-stakeholder, cross-sectoral partnerships between governments, the private sector, and civil society.

What further makes this study anomalous is the particular juncture in time at which it was implemented. Although background efforts, preparation, and prior consultations occurred in person, the first formal meeting of the Consensus Study Committee was held at the Mexican Academy of Sciences in Mexico City in March 2020, coinciding with the March 2020 declaration by the World Health Organization of COVID-19 as a global pandemic. In these circumstances, the drafting of the study design required flexibility and innovation, two of the very characteristics that, the study found, were essential for the sustainability partnerships on which it focused. The U.S. and Mexican Academies and the Committee recognized the implications of COVID-19 for safety, security, and mobility as well as the opportunities it posed. The process rapidly pivoted to a virtual study conducted remotely with a single webinar, organized to serve to inform committee deliberations as the empirical basis of the findings presented. This report would not have been possible without the input and commitment of stakeholders who shared their experiences at that July 2020 webinar. The committee anticipates that the extensive consultations they conducted exclusively online have enriched the deliberations and served to strengthen the binational and transdisciplinary nature of the report.

The Committee would also like to express appreciation to everyone who worked enthusiastically and tirelessly to craft this report, as well as the George and Cynthia Mitchell Endowment for Sustainability Sciences in collaboration with the Mexican Academy of Sciences, Academy of Engineering of Mexico, and National Academy of Medicine of Mexico for sponsoring this study. The report was greatly improved by the views, comments, and suggestions offered by the external reviewers. The committee is also indebted to the contributions of the Roundtable on Science and Technology for Sustainability. The committee expresses appreciation for the opportunity to work with José Luis Morán and Estela Susana Lizano Soberón from the Mexican Academy of Sciences, and gratefully acknowledges José Franco and Renata Villalba from the Mexican Academy of Sciences for their guidance and support. The committee is also thankful for the leadership and guidance provided by the project staff of the National Academies of Sciences, Engineering, and Medicine, including Jordyn White, the study director, along with Toby Warden, Adam Jones, and Daniel Talmage. Finally, we are grateful to our two consultants, Elisabeth Huber-Sannwald and Heather Kreidler.

Christopher A. Scott, *Chair*
Committee on Sustainability Partnerships in the
U.S.–Mexico Drylands Region

Contents

SUMMARY   1

1 INTRODUCTION   11
   Charge to the Committee, 12
   Sustainable Development Goal 17—Strengthening
      Global Partnerships, 13
   History of Collaboration Between the
      U.S. and Mexican National Academies on
      Binational Sustainability, 15
   Committee's Approach to the Study, 15
      A Focused Workshop Approach to a Consensus Study, 16
      Coordinating Stakeholder Engagement, 17
      Combining Research and Experiential Knowledge, 18
   Organization of the Report, 19
   References, 19

2 SUSTAINABILITY PARTNERSHIPS   21
   Partnerships, Sustainability Initiatives, and SDGs, 22
   Types of Sustainability Partnerships, 24
   How Partnerships Emerge, 26
   Characteristics of Partnerships for Sustainability, 28
      Trust, 28
      Participation, 29
      Coproduction of Knowledge, 30

Alignment, 31
Leadership, 32
Best Practices of Successful Partnerships, 33
Sustainability Partnership Persistence, 35
Principles of Effective Multi-Stakeholder Partnerships, 37
Summary, 39
Findings and Conclusions, 39
References, 40

3   **OPPORTUNITIES AND CHALLENGES FOR U.S.–MEXICO SUSTAINABILITY PARTNERSHIPS**   45
How Binational Partnerships Emerge and Evolve, 46
  Co-Creation and Capacity Building, 46
Indigenous Community Partnerships Across the
  U.S.–Mexico Border, 47
How Organizations Connect Around Sustainability Challenges, 49
Role of Information in Sustaining Partnerships, 51
The Effects of COVID-19 on Sustainability Partnerships, 52
What Successful Partnerships Look Like, 55
Webinar Summary, 56
Key Insights from the Webinar, 61

4   **RECOMMENDED STRATEGIES FOR EFFECTIVE PARTNERSHIPS**   63
The Socio-Ecological Systems Approach, 64
Recommended Strategies for Forming and Maintaining Successful
  U.S.–Mexico Binational Sustainability Partnerships, 66
Concluding Thoughts, 70
References, 71

APPENDIXES

A   STAKEHOLDER INFORMATION QUESTIONNAIRE   73
B   WEBINAR AGENDA   81
C   COMMITTEE MEMBER AND STAFF BIOGRAPHIES   85
D   CHARACTERISTICS OF THE BINATIONAL REGION   93
E   ACRONYM LIST   135

Summary

The border region shared by the United States and Mexico is currently experiencing multiple crises on both sides that present challenges to safeguarding the region's sustainable natural resources and to ensuring the livelihoods of its residents. These challenges are exacerbated by stressors including global climate change, increasing urbanization and industrialization and attendant air and water-quality degradation, and rapid population growth. Navigating these challenges and preserving the area's cultural richness, economy, and ecology will require building strategic partnerships that engage a broad range of stakeholders from both countries. Effective partnership strategies that support sustainable development can enhance both human well-being and interconnected ecological systems.

The U.S.–Mexico border states have maintained longstanding collaborations around water management, flood control, fire management, and the sharing of information and scientific findings related to the region's sustainability. However, as both countries' priorities for the region change (for Mexico, to serve as the gateway for binational commerce and foreign investment, and for the United States, increasingly as a buffer against immigration), it has become clear that additional innovative partnerships are needed among a diversity of agencies and organizations in the public, private, and civil sectors to foster comprehensive cross-border collaboration and the coproduction of regional solutions, interventions, and stewardship.

Building on a history of collaborative work on these and related opportunities, the U.S. National Academies of Sciences, Engineering, and Medicine, together with the Mexican Academy of Sciences (*Academia Mexicana de Ciencias*), Mexican Academy of Engineering (*Academia de Ingeniería de México*), and Mexican National Academy of Medicine (*Academia Nacional de Medicina de México*), appointed a committee of experts from the United States and Mexico to conduct a consensus study that identified partnership strategies to address select binational sustainability challenges.

This consensus report incorporates features of the United Nations 2030 Agenda for Sustainable Development, in particular, Sustainable Development Goal (SDG) 17. SDG 17 calls for revitalizing global partnerships for sustainable development. It is specifically focused on the advancement of multi-stakeholder partnerships that require coordination and collaboration among diverse stakeholders in pursuit of a common and mutually beneficial vision.[1] With attention to SDG 17, the report draws on social science theory and applied research on partnerships to explore potential strategies and mechanisms to increase coordination between relevant government agencies, the private sector (such as the mining and energy industries), and civil society in the United States and Mexico.

The committee defines U.S.–Mexico *binational sustainability partnerships* as:

> Organizations and individuals from different sectors and interest groups within the United States and Mexico, voluntarily coming together with organizations or individuals across the U.S.–Mexico border to address shared binational challenges and opportunities for sustainable development that isolated efforts or national initiatives would not be able to effectively accomplish.

To fully understand the state of partnerships in the region, the committee solicited input from stakeholders in the public and private sectors, government, academia, and civil society who are engaged in U.S.–Mexico binational partnerships. This stakeholder feedback, obtained via a questionnaire and a panel discussion at a July 2020 binational and bilingual webinar on sustainability partnerships, served to enrich the deliberations of the committee. The committee identified the following sustainability themes as the starting point for structuring the webinar (listed here alphabetically): Arts and culture, preservation; climate change and environmental conservation; critical resource management (water-energy-food); disaster and emergency management; education and research; environmental justice; humanitarian aid; migration; mining and extraction; public health; trade and commercial manufacturing; transportation; and urban planning and development. In addition to gathering input via the webinar, the committee assessed

---

[1] See: https://unstats.un.org/sdgs/report/2017/goal-17/.

the available scholarly literature on the characteristics of the U.S.–Mexico region and sustainability partnerships.

The report comprises four chapters. Chapter 1 provides an introduction to the statement of task and background on the committee process. Chapter 2 critically reviews the published literature and thinking on partnerships, placing it into context with the SDGs (both broadly and specifically to SDG 17) as well as with the characteristics of the binational region. Chapter 3 uses input from the July 2020 stakeholder webinar to explore key opportunities and challenges for sustainability partnerships. The final chapter outlines the committee's recommended strategies for effective partnership strategies. Appendix D reviews the binational context and characteristics of the region and gives context to binational partnership discussions elsewhere in the report.

The following concerns, drawing from prior collaborative work between the two academies,[2] were identified as priority areas by the committee for addressing sustainability challenges in the region: the increasing and evolving interactions and flows of people, resources, and services; a reconsideration of energy and industry based on the scarcity and abundance of natural resources; managing environmental and anthropogenic change in the midst of, and often resulting from, shocks and stressors, many of which are unique to the binational drylands region; and the benefit of governance and innovation that consider local communities and traditions while also keeping an eye on future challenges and opportunities for sustainable development.

CONCLUSIONS

Addressing these priority areas requires the fostering of strategic partnerships that engage a diverse set of stakeholders on either side of the border to devise strategies in support of sustainable development, by protecting the well-being of humans and ecosystems in the binational region. Based on the literature as well as input from the stakeholders, the committee concluded that successful partnerships require organizational flexibility, adaptation to change, financial resources, and norms of distribution, as well as the maintenance of an environment that fosters

---

[2] National Academies of Sciences, Engineering, and Medicine. 2018. *Advancing Sustainability of U.S.-Mexico Transboundary Drylands: Proceedings of a Workshop*. Washington, DC: The National Academies Press. https://doi.org/10.17226/25253; National Academies of Sciences, Engineering, and Medicine y Academia Mexicana de Ciencias, Academia de Ingeniería de México y Academia Nacional de Medicina de México. 2018. Avances en la Sostenibilidad de Tierras Áridas Transfronterizas de Estados Unidos y México. https://amc.edu.mx/amc/libros/drylands.

innovation, learning, collaboration, and trust. Knowledge co-production is key to combatting asymmetries and creating value in sustainability partnerships. Furthermore, it is imperative that stakeholders respect the knowledge and culture of the region by establishing informal community relationships and integrating Indigenous and local knowledge into the partnership strategies.

> CONCLUSION 1: The U.S.–Mexico border region faces many ongoing challenges in safeguarding the sustainability of its natural resources—scarce in some aspects yet abundant in others—to ensure the economic vitality and livelihoods of its people while protecting its cultural richness and unique natural environment.
>
> CONCLUSION 2: There is growing potential for partnership efforts around binational industrial, energy, and mining sustainability.
>
> CONCLUSION 3: Navigating the sustainability challenges in the U.S.–Mexico border region will require sound governance and the building and strengthening of strategic partnerships.
>
> CONCLUSION 4: Effective data sharing in transnational partnerships, or partnerships involving a mixture of private, public, and civil society actors with different sets of knowledge, experience, and information access, requires respecting the norms and institutional constraints of participants with enhanced transparency and accountability through partnership-specific data management protocols.
>
> CONCLUSION 5: Establishing informal community relationships and integrating Indigenous and local knowledge are instrumental in partnerships that span administrative levels and geographic boundaries.
>
> CONCLUSION 6: Knowledge co-production creates value in sustainability partnerships when it emanates from mutual or "horizontal" relationships among all the involved actors, confronting current power asymmetries with a commitment to combat inequality and exclusion.
>
> CONCLUSION 7: Partnership persistence requires a systemic approach toward a shared goal. It is a function of the partners' organizational flexibility, adaptation to change, financial resources, and norms of distribution, as well as whether they maintain an environment that fosters innovation, learning, collaboration, and trust.

CONCLUSION 8: Alignment as a process among partners to identify synergies for pursuing and securing the common good achieves coherent, efficient, and effective outcomes. Effective alignment requires flexibility in the partners' perspectives, values, and processes to enable coordination, identify appropriate entry points for new information integration, and achieve continuous learning.

SDG 17 acknowledges that "A successful sustainable development agenda requires partnerships between governments, the private sector, and civil society. These inclusive partnerships built upon principles and values, a shared vision, and shared goals that place people and the planet at the center, are needed at the global, regional, national and local level."[3] Ensuring that sufficient means of implementation exist to provide countries the opportunity to achieve the SDGs will require international cooperation; collaboration across the U.S.–Mexico Border is no exception. The committee agrees that partnerships can thrive or fail depending on partner norms concerning participation, relations of trust, transparency, and the acknowledgment of asymmetries in power, resources, and capacities. When there is a mix of actors from different sectors, the asymmetries in power can hinder the relationship building that is foundational to both participation in partnerships and their effectiveness. Complementarity in capacities and collaborative advantage is important to emerging partnerships.

ACHIEVING EFFECTIVE PARTNERSHIPS IN THE REGION

In this unique border region, the challenges involved in effectively pursuing sustainable development surpass the capacities of any single actor, type of actor, or government. Navigating these challenges and preserving the area's cultural richness, economy, and ecology will require strengthening existing—and building new—strategic partnerships that engage a broad range of binational stakeholders. Effective partnerships involve the application of knowledge and information, services, skills, financial resources, and engaging institutions, as well as an understanding of the expected outcomes.

In all sustainability contexts, environmental and social processes are tightly coupled. Multi-stakeholder partnerships targeting sustainable development in the U.S.–Mexico border region confront a complex cross-border socio-ecological system, one that requires both the ability to adapt and the ability to transform itself in response to a range of economic, cultural, political, social, and environmental interconnected dynamics. However, because there is a general lack of systematic data on existing partnerships

---

[3] For more information: https://unstats.un.org/sdgs/report/2017/goal-17/.

in the region, including data on the quality and effectiveness of those partnerships, they often proceed without a full understanding of the current social infrastructure or other paths through which sustainable development might be pursued—a consequence that can inhibit innovation and overall effectiveness.

Partnerships, particularly those aiming to address sustainability and sustainable development challenges in the region, and at the border, in particular, would improve effectiveness by adopting a complex social-ecological systems approach. This framework can help partners better understand the dynamics between the social and ecological processes to address problems through transformative, adaptive change. It would also help them effectively respond to unpredictable or extreme events such as the COVID-19 pandemic, severe drought, or other disasters. Aligning partnership strategies with the SDGs can also promote systems thinking and integrated development.

RECOMMENDED STRATEGIES FOR FORMING AND MAINTAINING SUCCESSFUL BINATIONAL SUSTAINABILITY EFFECTIVE PARTNERSHIPS

To be effective, binational sustainability partnerships must be centered on trust and have clear, mutually defined objectives, the ability to navigate power dynamics, transparency in partnership implementation (including flexibility, timing, and sequencing of activities), and, above all, process and governance as mechanisms to deliver partnership outcomes. The committee recommends six key strategies for forming and maintaining successful binational partnerships, as follows:

Strategy 1: Identify Critical Areas to Be Addressed by the Partnership

It is important for stakeholders to have a clear, mutual understanding of the explicit objectives of a partnership. Developing partnerships and understanding objectives involve identifying a target audience for activities and learning what impacts the partnership will have on other audiences and processes. When considering the desired outcome, partners also need to consider the assumptions around that outcome—such as resource availability and codependent processes—and the risks involved in pursuing it. Partners also need to identify tradeoffs and understand and accept that there is always uncertainty with respect to desired outcomes.

Strategy 2: Establish Trust Among Partners

Relationship building is essential to successful partnerships, often starting long before a formal partnership has been established among stakeholders and continuing long after it has ended. There is great value in practicing diplomacy within intergovernmental and civil society partnerships. However, a project's or a program's timing and a desire for efficiency often do not lend themselves to the patience and pace of learning societal norms and cultural sensitivity that help foster and build partners' trust.

In the case of partnerships among stakeholders from the United States and Mexico, particularly those involving representatives from local Indigenous communities, intercultural communication and competence—*interculturalidad*—is a key capacity. Developing new, beneficial relationships among stakeholders and actor groups involves establishing continuous open dialogue, having an agreed-upon partnership structure (often involving a formal memorandum of understanding), and the creation of a procedure for conflict resolution.

Strategy 3: Balance and Organize Power Dynamics

Achieving and maintaining successful multi-stakeholder partnerships requires the pursuit of "horizontal" interactions among partners that are fair and transparent. This means adopting a rotating leadership, even if the partners vary in size, organizational strength, financial standing, or other key characteristics. Addressing power asymmetries effectively requires active listening, such as that of academic engagement with Indigenous communities on the border, as well as awareness of the differential risks and responsibilities of engaging in partnerships for each actor. In these cases, equitability will arise from creating an operational plan for the partnership that factors in each stakeholder's organizational capacity and the complementarity of assets, as well as by ensuring the equal, equitable, and fair participation of actors in decision-making processes. It can also be helpful to view nontraditional attributes of strength and influence, such as social power, as equally enabling forces in partnership execution.

Strategy 4: Establish a Stable Governance Structure

Adopting strategies for effective partnerships requires a highly flexible and adaptive collaborative structure that incorporates robust decision-making and goal-oriented action. The overall approach requires strong leadership support to articulate and pursue short-, medium-, and long-term goals that set stakeholders' expectations for partnership effectiveness. Adaptive governance of multi-stakeholder partnerships entails the adoption of iterative approaches to monitoring, assessment, and interpretation of outexpectations, goals, projected impacts, and internal and external benefits of the partnership. Boyle and colleagues (2001) suggest that this type of transformative governance is the process of continuously targeting the collective benefits (and values) while responding to and resolving tradeoffs in the pursuit of sustainable development.

The complex sustainability context in the U.S.–Mexico cross-border region may cause governance gaps, in which stakeholders confound challenges with actors (Bergsten et al., 2019), attributing responsibility for certain outcomes to institutions or individuals who may have little control over the circumstances. Open communication, sharing of analogous experiences, and collaborative identification of responses can mitigate these situations.

Strategy 5: Agree on a Definition of Effective Partnership Execution

For partnerships to succeed there needs to be a clearly defined outcome and a mutual commitment by each partner to execute the outcome. Although there are numerous similarities between cities and industries in the U.S.–Mexico border region, each country and each stakeholder group's conceptualization of partnership success is likely to vary—and they may at times contradict one another. While desired outcomes may evolve, mutual commitment and a trusted process can ensure that such evolution brings all partners forward in continued collaboration. Adding guidelines for partner compliance and using tools that can aid in practical decision making can help validate the partnership process and legitimize the partnership. Each actor should demonstrate a continued commitment to engaging in and achieving partnership goals.

Strategy 6: Develop Short-, Medium-, and Long-Term Goals

Partnership strategies can be applied over different timeframes. While sustainable development is a long-term goal, pursuing it requires consistent short- and medium-term efforts, which will be enhanced through partnership-based initiatives of the kind detailed in this report. Such a strategy goes beyond recognizing that different timeframes apply to different goals, and hence that some are to be pursued in the short, medium, or long term. Consideration of timing, as pertaining to partnership strategy, has to evaluate the sequence of tasks so that each activity maximizes the probability of achieving the aims of the next step. This is crucial in the context of reaching broader SDGs, for these can only be attained by building on necessary preconditions. Effective partnerships require a strategy that is mindful of the timing and sequence of the assumed tasks, since appropriate timing and sequence are crucial for reaching SDGs.

Strategy 7: Establish Guidelines for Partnership Evaluation

There are three key measures for assessing partnerships: *process* (partnership formation, goal setting, defining stakeholder roles, and conducting partnership activities); *governance* (flexibility, equity, accountability, responsiveness, transparency, and consistency among partners and external stakeholders); and *outcomes* (results in relation to goals and associated tangible factors that emerge from partnership activities). The criteria of effective process, governance, and outcomes are interwoven with principles for sound partnerships, chiefly, principles to guide institutional transformation, social and political power, conflict, communication, and leadership.

Process guidelines for effective partnerships start with the way clear goals are achieved, with participants and external stakeholders jointly defining the roles and responsibilities they will pursue, and where appropriate, modifying goals. Both formal and informal means of participation are important, though each must be understood, monitored, and promoted distinctly. For example, in a pandemic, informal participation may gain temporary priority. It is essential for partnership participants and leaders to be aware of, and seek to promote, equity through procedural justice to incorporate and address the needs of less dominant actors and groups. Latent and overt forms of internal conflict can destabilize both emerging and established partnerships if not harnessed as a force for positive change, for example, when legal pursuits by Indigenous communities are used to assert resource rights. The choice of leadership approaches and the establishment of checks and balances are critically important, in process terms, when leaders are themselves involved in, or may be the cause for, conflict. These final two process guidelines—navigating power and conflict—are ultimately also governance challenges.

Governance guidelines include flexibility and responsiveness, especially the ability to produce qualitatively different strategies for different approaches to partnership goals, activities, and outcomes. Co-production of knowledge and process within partnerships (among members and leadership) and for partnerships with external stakeholders or constituents influence the quality of those partnerships, the initiatives they pursue, and the broader communities of practice they build and sustain. Additional governance guidelines for partnerships involve setting and maintaining policies and procedures, including (where necessary) legal agreements, that enhance transparency and predictability as well as improve and ensure coherence of policy and institutional aims.

Outcome guidelines for a partnership, that is, the degree to which results and impacts are generated, sustained, and equitable, are perhaps the best signal to external constituents that partnerships are effective. Given the focus of this study on SDG 17, a more nuanced appreciation of local needs and context-specific indicators of the suite of SDGs is an important consideration. For example, binational water-management partnerships are crucially important to enhance water security in this arid and semi-arid region, which is confronting growing water demands for human and ecosystem needs. Additional, key considerations for partnership outcomes include resources, both material and financial, as well as capacities. Partnerships' abilities to mobilize and deliver such outcomes as knowledge sharing, expertise, technologies, and financial resources are central to their pursuit of achieving sustainable development locally, in the binational region, and globally.

1

Introduction

The U.S.–Mexico border region currently faces multiple sustainability challenges at the intersection of the human and natural systems affecting both nations. Warming and drying conditions are threatening surface water and groundwater availability, disrupting farming, grazing, and other land- and marine-based livelihood systems, and challenging the sustainability of human settlements and economic activity in the region. These biophysical challenges are exacerbated by a highly mobile and dynamic population, insecurity, poverty, volatile economic conditions, an often-tense border-policy environment, increased exposure to extreme weather events, and urbanization on marginalized lands. In short, social and political processes are inextricably linked to ecological dynamics in this border social-ecological system.

There has been a long history of collaborations among U.S.–Mexico border states pertaining to water resource management, flood control, fire management, and information exchange associated with climate variability and change impacts (National Academies of Science, Engineering, and Medicine [NASEM], 2018; Wilder et al., 2020). These collaborations have often arisen despite the dynamic context of border policy at the national level in the two countries, in which trade asymmetries, cross-border migration, the management of illegal commerce, and natural resource management challenges have often created tensions in bilateral relations. There is growing awareness of shared social and ecological challenges and potential responses and that significantly more remains to be done to develop the binational scientific, policy, and management capacity that is needed to promote sustainable development. It is of increasing importance to advance innovative partnerships among a diversity of public, private, and civil sectors that strengthen comprehensive cross-border collaboration and the co-production of sustainability solutions, interventions, and knowledge.

Understanding what makes some partnerships succeed while others fail or falter requires looking to both social science theory and practice on the ground. Knowledge from the social sciences, such as theories related to social change, managing transdisciplinary initiatives, social-ecological governance models, and participatory action research yields insights into effective partnership structures and strategies (Stibbe et al., 2019). Lessons learned from successful case studies are also valuable for understanding how partnerships work *in situ*. Furthermore, to determine what types of partnerships are required for success in the region, a better understanding is needed of the complexity of the challenges that cross varying scales, geographic regions, and financial constraints (Lutz-Ley et al., 2020).

CHARGE TO THE COMMITTEE

To better understand these challenges, the National Academies of Sciences, Engineering, and Medicine, with support from the George C. Mitchell Endowment for the Sustainability Sciences and in collaboration with the Mexican Academy of Sciences, Academy of Engineering, and National Academy of Medicine, undertook a study to identify actionable approaches to advance the efficacy of partnerships for sustainability in the drylands border region shared by the two countries. This consensus activity combined the thematic and regional expertise of committee members with insightful and often challenging views shared by a diverse group of stakeholders from across the public, private, and civil society sectors during a structured webinar. The committee included experts in the areas of sustainability, social change theories, drought and water resource management, institutional capacity building, policy and regulatory decision making, and environmental change, as well as individuals with industry and practitioner experience and expertise.

Committee and stakeholder discussions centered on partnership strategies for sustainable development and were supported by a thorough review of literature on partnerships and literature on the border region's biodiversity and social-ecological systems. The objective of the webinar was to inform committee deliberations and in turn to enhance future collaborative efforts focused on putting knowledge into action. Box 1-1 contains the full statement of task for the committee.

INTRODUCTION

> **Box 1-1**
> **Statement of Task**
>
> The U.S. National Academies of Sciences, Engineering, and Medicine, jointly with the Mexican Academy of Sciences, Academy of Engineering, and National Academy of Medicine, will appoint a binational, ad hoc committee of experts from the United States and Mexico to identify partnership approaches to address select sustainability challenges in the binational drylands region outlined in the National Academies of Sciences, Engineering, and Medicine's 2018 workshop report, Advancing Sustainability of U.S.–Mexico Transboundary Drylands. The committee's work will support efforts identified through Goal 17 (partnerships for the goals) of the United Nations Sustainable Development Goals (SDGs), which calls for multi-stakeholder, cross-sector partnerships between governments, the private sector, and civil society.
>
> The committee will draw on social science theory and research around partnerships, and will review relevant case studies to explore short-, medium-, and long-term strategies and mechanisms to increase coordination between relevant government agencies, the private sector (such as the mining and energy industries), and civil society in the U.S. and Mexico. Through consultation with stakeholders, the committee will recommend potential strategies to address specific targets within the relevant sustainable development goals and determine the appropriate scale (local, regional and national) and timeframe for partnerships to align with the targets.
>
> Recommended strategies will consider social and environmental compatibility and the opportunity for economic growth.

SUSTAINABLE DEVELOPMENT GOAL 17— STRENGTHENING GLOBAL PARTNERSHIPS

The effects of climate change, air and water pollution, resource overconsumption, human migration, international trade (both formal and illicit), and a host of other social and ecological pressures are acutely felt in communities in the U.S.–Mexico binational region. The region exemplifies the dynamics of nested and interacting complex social-ecological systems in that social processes such as urbanization, migration, resource extraction, and trade constantly produce changes in the biophysical environment while being directly affected by environmental change. Developing solutions to these sustainability issues requires engagement and collaboration across societal sectors with attention to this dynamic coupling of society and the environment. In this context, both countries are rapidly increasing their capacity to understand climate-related challenges and opportunities. But significantly more needs to be done to develop the binational scientific, policy, and management capacity that is needed to promote sustainable development. Residents and other actors within the region have had relatively less capacity and political influence to shape such national policies to their benefit. Many challenges occur across jurisdictional boundaries and require resources beyond the capabilities of individual sectors to resolve. In advancing Sustainable Development Goals (SDGs) in this region, the stakeholders include governmental, tribal, industry, academic, local community, and nongovernmental actors (Alejo, 2019).

The United Nations sets forth in its 2030 agenda 17 SDGs (United Nations, General Assembly, 2015). Mexico as a nation, and in particular the Mexican federal government, committed actively and early to the SDGs (Lucatello, 2015; Mexico National Council for the 2030 Agenda for Sustainable Development, 2018; Ulfgard, 2017). Of the goals, the committee is tasked with furthering the work to achieve SDG 17,[1] which is to "Strengthen the means of implementation and revitalize the Global Partnership for Sustainable Development." Of that goal's 19 individual targets, the committee has identified targets 16 and 17 (below) as being particularly relevant for improving partnerships in the U.S.–Mexico binational region.

Target 17.16   Enhance the Global Partnership for Sustainable Development, complemented by multi-stakeholder partnerships that mobilize and share knowledge, expertise, technology, and financial resources, to support the achievement of the sustainable development goals in all countries, in particular developing countries.

Particularly with Target 17.16, the committee emphasizes, in both conceptual and programmatic terms, multi-stakeholder partnerships (as described in Chapter 2) with empirical case evidence (described in Chapter 3).

Target 17.17   Encourage and promote effective public, public-private, and civil society partnerships, building on the experience and resourcing strategies of partnerships.

Target 17.17 dovetails with the task of the committee, which is to better understand the opportunities and challenges for sustainability partnerships in the binational region to synthesize recommended strategies. These strategies are outlined in Chapter 4.

---

[1] More information about SDG 17 is available at: https://sdgs.un.org/goals/goal17.

HISTORY OF COLLABORATION BETWEEN THE U.S. AND MEXICAN NATIONAL ACADEMIES ON BINATIONAL SUSTAINABILITY

In May 2018, the U.S. National Academies of Sciences, Engineering, and Medicine and the Mexican Academy of Sciences, Academy of Engineering, and National Academy of Medicine[2] held a binational workshop. The outcome of that workshop was published under the title, *Advancing Sustainability of U.S.–Mexico Transboundary Drylands: Proceedings of a Workshop* (NASEM, 2018), in both English and Spanish (*Avances en la Sostenibilidad de Tierras Áridas Transfronterizas de Estados Unidos y México*). The proceedings highlighted the key sustainability challenges facing the region, explored the scientific and technical capacity that each nation can bring to help address them, and suggested new opportunities for binational research collaboration and coordinated management in the advancement of sustainability science and development (NASEM, 2018).

The workshop was centered around four sustainability themes of high priority to the binational region. The four sessions covered (1) the interaction and flow of resources, people, and services across the border and throughout the region; (2) the simultaneous scarcity and abundance of cultural and ecological resources; (3) environmental shocks and stressors, which often co-occur alongside unexpected policy changes and market volatility; and (4) how sustainable solutions can be achieved through governance and innovation at the local, national, and binational levels. These four themes tie the discussion of sustainability partnerships in this report more closely to the U.S.–Mexico binational region. The themes also served as guidance for this study committee as it carried out its deliberations.

COMMITTEE'S APPROACH TO THE STUDY

The consensus study committee comprises 11 experts, with representation from both the United States and Mexico, in the following disciplines: sustainability science; water resources management; social change and social justice; drylands ecology; policy making and institutions; climate and environmental change; and mining and industrial development.

At the onset of its work, the committee deemed it essential to develop a single, streamlined definition of binational sustainability partnerships, specifically directed toward U.S.–Mexico border relations. In defining sustainability partnerships, this committee builds on a definition that appeared in a National Research Council (2009) workshop summary (*Enhancing the*

---

[2] The *Academia Mexicana de Ciencias*, the *Academia de Ingeniería de México*, and the *Academia Nacional de Medicina de México*, respectively.

> **BOX 1-2**
>
> **Committee Definition of US Mexico Binational Sustainability Partnerships**
>
> Organizations and individuals from different sectors and interest groups within the United States and Mexico, voluntarily coming together with organizations or individuals across the United States–Mexico border to address shared binational challenges and opportunities for sustainable development that isolated efforts or national initiatives would not be able to effectively accomplish.

Effectiveness of Sustainability Partnerships), which defined them as "actors from different sectors (thereby excluding cooperation within a sector; e.g., business to business) voluntarily coming together to jointly produce what no single actor could effectively produce on its own" (p. 3). The committee's expanded definition appears in Box 1-2.

As described, such partnerships are especially beneficial for addressing challenges that call for cross-sectoral, interdisciplinary, collaborative solutions.

A Focused Workshop Approach to a Consensus Study

In addition to building on a prior binational collaborative workshop report, reviewing the literature on binational partnerships, and drawing on committee member expertise, the committee also sought key input through stakeholder feedback. It did this by conducting a public webinar on U.S.–Mexico binational sustainability partnerships and using the feedback to inform committee deliberations.

The consensus study design was centered on a focused workshop approach, in which committee members actively engage with other participants to discuss and obtain insights on key issues to be addressed in the statement of task. This approach jumpstarts the consensus process by having a committee plan and participate in a highly structured public workshop—the discussions from which serve as the primary information-gathering source for later committee deliberations in closed session. Due to restrictions in response to COVID-19, the workshop was held as a webinar. Because this activity was designed to inform a wide range of stakeholders in the region, gathering input and feedback from the various organizations and individuals that work on sustainability at the border and the binational region more broadly proved to be insightful while enabling the overall study process to be more inclusive and collaborative.

Coordinating Stakeholder Engagement

To develop the agenda for the workshop and a plan for the report, the committee first created a list of key sustainability themes that were most relevant to the environment, commerce, and culture in the U.S.–Mexico region, basing the list on the takeaways from the May 2018 workshop as well as the committee's existing knowledge of the region. Committee members then mapped the themes onto the four contexts identified at the May 2018 workshop—namely, interactions and flows, scarcity and abundance, shocks and stressors, and governance and innovation—and prioritized the list of themes based on their relevance to the four contexts. The committee ultimately settled on the following list of priorities, listed here alphabetically:

- Arts/Culture/Preservation
- Climate Change/Environmental Conservation
- Critical Resource Management (Water/Energy/Food)
- Disaster/Emergency Management
- Education/Research
- Environmental Justice
- Humanitarian Aid
- Migration
- Mining/Extraction
- Public Health
- Trade/Commercial Manufacturing
- Transportation
- Urban Planning and Development

In addition to the above themes, and as called for in the statement of task, the webinar and report were informed by the United Nation SDG framework, with a particular emphasis on partnerships.

To generate a broad list of attendees for the webinar and hear from a variety of stakeholders in the region, early in the study process the committee developed an online questionnaire to assess the landscape of partnerships between the United States and Mexico. The questionnaire, provided in English and Spanish, asked respondents to identify the sectors in which they conduct business and partnership affairs, the sectors in which their partner(s) operate(s), and their assessment of the effectiveness of these partnerships. The complete questionnaire text is reproduced in Appendix A.

A link to the online questionnaire was posted on the committee's website and was also distributed by the committee and staff to their binational and sustainability networks, with a request that contacts share the link with anyone they knew with an interest in binational partnership activity in the region. The questionnaire responses themselves were not used as data, nor were they a representative sample of all partnerships in the region; rather, they were used solely to help generate a broad and diverse list of invitees to participate in panel discussions. In total, 124 responses were received from governmental and nongovernmental organizations with partners involved in the aforementioned key sustainability areas. Over half of all responses were from stakeholders in Mexico. The committee thoroughly reviewed the responses and aggregated them by sector, and then sought to invite from each of the sectors represented in the responses at least one stakeholder to be a panelist in the webinar. Considering criteria such as diversity and inclusion, binational representation, and respondents' self-assessment of partnership history and effectiveness, the committee selected representatives of the following agencies and organizations to serve as webinar panelists:

- Arizona State University
- Consejo Empresarial Nogales A.C.
- El Colegio de la Frontera Norte
- Index Nogales, Asociación de Maquiladoras de Sonora, A.C.
- Líderes Tradicionales de O'odham in México
- Next Generation Sonoran Desert Researchers (N-Gen)
- Northern Arizona University
- San Diego Association of Governments
- U.S. Geological Survey
- U.S.–Mexico Border Philanthropy Partnership
- Watershed Management Group

The committee held the half-day, public, virtual stakeholder webinar on July 15, 2020. In the webinar, which was publicly broadcast, participants engaged in panel discussions moderated by consensus study committee members (see full agenda in Appendix B). Sessions were conducted in English and Spanish with bilingual translation available throughout the webinar. The webinar was recorded and transcribed and is discussed in detail in Chapter 3 of this report.

Combining Research and Experiential Knowledge

Following the workshop, the committee members met virtually in closed session several times, with writing teams meeting on an ad hoc basis, to discuss their charge. The committee then determined how to ground the webinar discussions in the context of its emerging thinking and reached a consensus in identifying strategies to enhance sustainability partnerships in the U.S.–Mexico binational region.

Having reviewed the literature on the dynamic climate, population, commerce, and natural resource characteristics of the region, and looking at research on effective partnerships, particularly in the context of SDG 17, the committee then sought to place the information received at the webinar into the larger partnership narrative, considering both countries' policies on trade, migration, the environment, and scientific cooperation. In this way, webinar-based insights on coordination among relevant government agencies, the private sector, and civil society were included in the partnership narrative along with an analysis of the notable strengths and weaknesses of each partnership type.

ORGANIZATION OF THE REPORT

This report comprises four chapters. Chapter 2 critically reviews the published literature and analyses on partnerships, placing it in context with the SDGs (both broadly and specifically to SDG 17), as well as with the characteristics of the binational region. Chapter 3 uses evidence from the July 15, 2020, webinar (see Appendix C for the webinar agenda) to explore key opportunities and challenges for sustainability partnerships. The final chapter outlines the committee's recommended strategies for effective partnership strategies. Appendix D reviews the binational context and characteristics of the region and gives context to binational partnership discussions elsewhere in the report.

REFERENCES

Alejo, A. (2019). Contemporary diplomacy, global politics, and nongovernmental actors: Dilemmas of the multistakeholder mechanism of participation in Mexico. *Politics & Policy, 47*(1), 105–126.

Lucatello, S. (2015). México y la agenda mundial del medioambiente en el escenario posterior a 2015. *Revista Mexicana de Política Exterior, 103*, 189–207.

Lutz-Ley, A.N., Scott, C.A., Varady, R.G., Ocampo, A., Lara-Valencia, F., Buechler, S.J., Díaz Caravantes, R., Zuniga Teran, A., Wilder, M.O., Martín, F., and Ribeiro Neto, A. (2020). Dialogic science-policy networks for water security governance in the arid Americas. *Environmental Development*. doi: 10.1016/j.envdev.2020.100568.

Mexico National Council for the 2030 Agenda for Sustainable Development. (2018). *Informe Nacional Voluntario para el Foro Político de Alto Nivel sobre Desarrollo Sostenible: Bases y fundamentos en México para una visión del desarrollo sostenible a largo plazo*. Mexico: Gobierno de la República. Avaliable: https://www.gob.mx/agenda2030/documentos/informe-nacional-voluntario-para-el-foro-politico-de-alto-nivel-sobre-desarrollo-sostenible-espanol.

NASEM (National Academies of Sciences, Engineering, and Medicine). (2018). *Advancing Sustainability of U.S.–Mexico Transboundary Drylands: Proceedings of a Workshop*. Washington, DC: The National Academies Press. doi: 10.17226/25253.

National Research Council. (2009). *Enhancing the Effectiveness of Sustainability Partnerships: Summary of a Partnership*. Washington, DC: The National Academies Press. doi: 10.17226/25253.

Stibbe, D.T., Reid, S., and Gilbert, J. (2019). *Maximising the Impact of Partnerships for the SDGs: A Practical Guide to Partnership Value Creation*. Oxford, UK: The Partnering Initiative and United Nations Department of Economic and Social Affairs (UNDESA).

Ulfgard, R.V. (Ed.). (2017). Mexico and the post-2015 development agenda: Contributions and challenges. In *Governance, Development, and Social Inclusion in Latin America*. New York, NY: Springer.

United Nations, General Assembly (2015). *Transforming Our World: the 2030 Agenda for Sustainable Development*. Available: https://www.un.org/ga/search/view_doc.asp?symbol=A/RES/70/1&Lang=E.

Wilder, M.O., Varady, R.G., Gerlak, A.K., Mumme, S.P., Flessa, K.W., Zuniga-Teran, A.A. Scott, C.A., Pineda Pablos, N., and Megdal, S.B. (2020). Hydrodiplomacy and adaptive governance at the US–Mexico border: 75 years of tradition and innovation in transboundary water management. *Environmental Science & Policy, 112*(October), 189–202.

2

Sustainability Partnerships

This chapter draws on social science and research related to sustainability partnerships, with attention to a broad global context, to lay a conceptual foundation for understanding the partnership efforts in the U.S.–Mexico binational region. The literature that informs this chapter is drawn from international research on multi-stakeholder and multinational partnerships addressing sustainability challenges, research on transboundary/multinational water and natural resource management partnerships, and literature on multi-stakeholder sustainability initiatives.

Following a section that covers definitions of partnerships in relation to sustainability initiatives and the U.N. Sustainable Development Goal (SDG) 17, the chapter addresses what is known about types of partnerships, the emergence of partnerships, common characteristics of partnerships, partnership governance, and, more normatively, what experts perceive as principles for effective sustainability partnerships. These themes, particularly in relation to identifying the characteristics of sustainability partnerships, are drawn on in Chapters 3 and 4 for greater insight into the specific sustainability partnerships of the U.S.–Mexico border region. The material of this chapter may also be useful for organizations and groups seeking to improve their partnership activities in light of sustainability goals and those who may be interested in common attributes and structures of similar partnerships internationally. While many themes and issues in the international literature do resonate with the challenges and structure of sustainability partnerships in the U.S.–Mexico region, the region also has unique features that create specific opportunities and obstacles to partnership initiatives. These features are discussed in Chapter 4 alongside a broader rationale for the application of a social-ecological systems framework.

PARTNERSHIPS, SUSTAINABILITY INITIATIVES, AND SDGS

"Collaboration across societal sectors," write Stibbe, Reid, and Gilbert (2019), "has emerged as one of the defining concepts of international development in the 21st century" (p. 6). Partnerships are now considered essential to sustainable development and the achievement of the SDGs. In the mid-1990s, the definition of sustainability partnerships were defined as voluntary collaborations between two or more organizations with a jointly defined agenda focused on a discrete, attainable, and potentially measurable goal (Long and Arnold, 1995). More recently in relation to the SDGs, the United Nations has adopted the following definition of sustainability partnerships:

> Multistakeholder initiatives, voluntarily undertaken by governments, inter-governmental organizations, major groups, and other stakeholders, which efforts are contributing to the implementation of inter-governmentally, agreed on development goals and commitments. (Stibbe et al., 2019, p. 8)

Partnerships are the specific focus of SDG 17, which encourages and promotes different stakeholders in the private and public sectors and civil society to collaborate in the achievement of the SDGs by pooling financial resources, technologies, knowledge, and expertise. These types of partnerships represent a critical means of implementing the whole sustainability agenda and achieving all the SDGs.[1] As it pertains to achieving the SDGs, a multistakeholder partnership is defined as:

> an ongoing collaborative relationship among organizations from different stakeholder types aligning their interests around a common vision, combining their complementary resources and competencies and sharing risk, to maximize value creation towards the Sustainable Development Goals and deliver benefit to each of the partners. (Stibbe and Prescott, 2020, p. 6)

SDG 17 acknowledges that:

> [a] successful sustainable development agenda requires inclusive partnerships—tthe global, regional, national and local levels—built upon principles and values, and upon a shared vision and shared goals placing people and the planet at the centre.[2]

---

[1] More information is available at: https://sustainabledevelopment.un.org/sdinaction and https://www.un.org/sustainabledevelopment/globalpartnerships/.

[2] More information is available at: https://www.un.org/sustainabledevelopment/globalpartnerships/.

SDG 17 and its respective targets, particularly targets 17.16 and 17.17 (see Chapter 1), identify multi-stakeholder partnerships as essential to mobilize and share information, knowledge, technologies, and financial resources to achieve sustainable development worldwide, particularly in developing countries. SDG Target 17.17 seeks to "encourage and promote effective public, public-private and civil society partnerships, building on the experience and resourcing strategies of partnerships" (United Nations, 2015). Increased knowledge sharing and access to technology are key ways to distribute information and encourage innovation.

The latest U.N. (2020) report on the progress of SDG 17 indicates that the "financial resources remain scarce, trade tensions have been increasing, and crucial data are still lacking" and that "[s]trengthening multilateralism and global partnership are more important than ever" (United Nations, 2020, p. 58). The COVID-19 pandemic is threatening trade and foreign direct investment. Major donors will strive to protect official development assistance (ODA)[3] budgets while worldwide remittances in 2020 are estimated to decrease by approximately 20 percent—the largest decrease in recent history. Additionally, in 2020 the receipt of global foreign direct investment by developing countries may decrease by up to 40 percent due to postponed investments. Furthermore, global merchandise trade is estimated to decrease by 13 to 32 percent. While the pandemic has forced many people to rely on the Internet, almost half of the world's population—concentrated in poorer countries—is not connected.

Finally, it is reported that while the need for sound data and statistics has increased, many countries lack the necessary technical and financial resources for monitoring development agendas (United Nations, 2020). In sum, the need persists for partnerships to bridge these gaps in finance, information, and commerce. Partnerships—between nations and between public, private, and civil society entities—are considered vehicles for helping accomplish these goals (Prescott and Stibbe, 2020; Stibbe and Prescott, 2020). The development of multi-stakeholder partnership platforms throughout the world has the potential to hasten steps forward toward achieving the SDGs. In general terms, these platforms have four objectives: (1) joint advocacy and policy dialogue to create an enabling environment where partnership thrives; (2) partnering at scale for impact: support to identify large-scale public-private partnerships and collaborations; (3) maximizing innovative finance; and (4) facilitating data management, learning, and research to inform progressive policy and practice for SDG partnerships (Prescott and Stibbe, 2020, p. 18).

---

[3] More information on official development assistance (ODA) is available at: https://www.oecd.org/dac/financing-sustainable-development/development-finance-standards/officialdevelopment assistancedefinitionandcoverage.htm.

To ensure that countries have the opportunity to achieve the SDGs will require international cooperation; collaboration across the U.S.–Mexico border is no exception. Ongoing climate change, land degradation, social instability, and other binational challenges make achieving the SDGs in the U.S.–Mexican transboundary region both daunting and urgent. Multi-stakeholder partnerships for sustainable development aim at integrating various sectoral and disciplinary perspectives on a broad spectrum of essential needs, including food (SDG 2), water (SDG 6), energy (SDG 7), ocean resources (SDG 14), and terrestrial ecosystems (SDG 15), which are fundamental to achieving the expected objectives for all to live poverty-free (SDG 1), healthy (SDG 3), with access to quality education (SDG 4), securing gender equality (SDG 5), with unconstrained access to labor and economic rights (SDG 8), securing social equality (SDG 10), and in an overall inclusive society (SDG 16). To meet the essential needs and expected objectives, science needs to become policy-relevant, and novel governance structures should secure resilient infrastructure (SDG 9), sustainable cities and communities (SDG 11), responsible production and consumption schemes (SDG 12), and effective climate-change mitigation action (SDG 13) (Fu et al., 2019). (Appendix D offers discussion of many of these topics in the context of the U.S.–Mexico transboundary region.)

TYPES OF SUSTAINABILITY PARTNERSHIPS

In different analyses in the research literature, multi-stakeholder partnerships are categorized by their functions, by their aim or scope of action, by the type or organizational level of actors involved, or by their degree of temporal permanence and formal institutionalization (for a few examples, see Gurzawska, 2020; Pinkse and Kolk, 2012; van Huijstee and Glasbergen, 2010). Much of the analysis of international or transnational multi-stakeholder partnerships arise from political science and international relations, rather than directly from the field of sustainability. Schäferhoff, Campe, and Kaan (2009), for example, classify transnational partnerships according to whether they are primarily dedicated to policy formation, such as the development of common norms or standards, or whether they are more focused on policy implementation. These two general purposes of partnerships can be further classified in terms of their primary functions: advocacy, awareness-raising, service provisioning, knowledge exchange, research and development, standard setting, or the creation of markets.

Focusing on public-private partnerships in transnational governance, Börzel and Risse (2002) describe a similar typology of partnerships according to purpose: those involved primarily in rule formation, those dedicated to rule implementation, and those focused on service provisioning. In relation specifically to the accountability of partnerships in climate action, Bäckstrand (2008) builds on Börzel and Risse's typology and relates partnership functions and accountability to the nature of participating entities: public-private, government-government, and private-private.

Garrick et al. (2018) classify partnerships in terms of scope, and subsequently by authority level and formality:

- *Single issue:* These informal policy networks materialize out of local interactions for common ventures or from service contracts to deal with externalities, such as dry-year options in water management partnerships.
- *Multilateral:* These are multipurpose partnerships with a consolidated set of public services within the geographic territory; for example, regional or watershed organizations that coordinate drought response in a watershed.
- *Comprehensive:* These are regional partnerships with embedded norms formed as a result of intersecting ventures, agreements, contracts, and coordination throughout numerous policy domains, controlled by a statutory framework; for example, water quality planning by a joint river basin authority.

Typically, when collective action benefits exceed costs, decision-making venues increase in scope and authority. This suggests that informal venues may be sufficient until the capacity of the partners is exceeded. Lower-cost integration mechanisms are likely used and experimented with before more comprehensive and formal mechanisms (Garrick et al., 2018).

With specific reference to sustainability and sustainable development, multi-stakeholder partnerships have often been classified according to the intent of their collective action and the duration of their collaborations. For example, Peterson and colleagues (2015) identify four types of partnerships: (1) joint projects, focused on a short-term, one-time collaborative effort; (2) joint programs, representing a small number of partners working on an explicit portion of a social problem; (3) strategic alliances, in which partners create platforms for ongoing collaboration to tackle one or more related social issues supporting a common agenda and investments; and (4) collective impact partnerships, partnerships designed for long-term commitments to a common agenda by cross-sector actors aiming for systemwide change.

In the U.S.–Mexico region, all four of these forms of partnerships are likely. For sustained impact on the persistent sustainability challenges of the region, such as the concerns related to migration, water resource management, and public health, strategic alliances and collective impact partnerships may be particularly constructive. Chapter 3 covers some current partnerships, such as the work of the Border Philanthropy Partnership, which could be considered a partnership striving for longer-term collective impact by facilitating financial resource access for actors in diverse sectors, addressing diverse sustainability challenges in the border region.

The San Diego Association of Governments, as a forum for decision making in the broader San Diego region, has characteristics that echo a strategic alliance partnership. Other partnerships—for example, collaborations among universities on both sides of the border to address educational or research objectives—may begin as joint projects that subsequently evolve into more programmatic partnerships or even into broader alliances. The capacity for informal and formal joint projects and programs to emerge to address abrupt concerns, such as the COVID-19 crisis, may also depend on the social and institutional infrastructure of existing strategic alliances.

The Alianza Indígena Sin Fronteras exemplifies all of these functions of partnerships, mobilizing and strengthening in response to threats to the diverse cultural and environmental resources stemming from the fortification of the U.S.–Mexico border. This action-oriented partnership is also serving to communicate ideas and knowledge among border research institutes and Native communities, while serving as a community-based partnership, reinforcing ties among native peoples across the border region. Research-based partnerships, particularly those involving academics and professionals in public resource management institutions, have long been involved in cross-border collaborations, addressing the sustainability challenges in water resources, biodiversity, and natural hazards domains. Communities of practice have also emerged, focusing on key sustainability concerns such as cross-border migration and trade.

HOW PARTNERSHIPS EMERGE

Scholars have highlighted the emergence and proliferation of multi-stakeholder transnational partnerships as a form of governance in the neoliberal era. These forms of governance have responded to, on one hand, perceived market and state failures in access to critical goods and services (Pattberg and Widerberg, 2014). On the other hand, they generate novel opportunities for private-sector and civil society participation and influence in decision making, as public-sector actors have retreated from some obligations providing for the public good (Börzel and Risse, 2002; Scäferhoff et al., 2009).

Management of sustainability concerns across political boundaries is particularly challenging. For example, concerning transborder water management, Ingram, Milich, and Varady (1994) identify five potential difficulties (1) political boundaries (domestic or international) can separate a location where a problem is felt from the effective and efficient solutions; (2) economic opportunities for profit can make the moderation of scarce-resource use unlikely; (3) borders can exacerbate perceived inequalities; (4) residents' concerns can be marginalized; and (5) policies can impede grassroots problem-solving. They further note that policies at the federal and state level are often at odds with the needs and priorities of the region. In such conditions, partnerships can emerge and play critical roles. Furthermore, state actors can encourage the formation of transnational partnerships as a means of enhancing the legitimacy and effectiveness of transnational policy initiatives to address complex and difficult problems (Börzel and Risse, 2002), as well as a means of dispersing responsibility and risk among a broader coalition of actors.

Sustainability partnerships are often thought to emerge in contexts where an organization recognizes the added value of working with others toward aligned goals or within a common agenda. In some cases, the organization may perceive that its agenda and goals cannot be easily met without the contributions of other actors or organizations: that is, they might identify a "collaborative advantage," which Stibbe, Reid, and Gilbert (2019) define as "the alchemy that allows a group of actors to collectively deliver more than the sum of their input parts" (p. 11). In this case, partnerships may be formed featuring organizations that provide complementary skills, relationships, resources, or other critical assets (NRC, 2009; Schäferhoff et al., 2009). Partnerships can emerge when participants recognize that collaboration is a means of access to skills, resources, network funding, or influence that they might not otherwise have (Schäferhoff et al., 2009).

In other cases, an organization may see that its goals, while distinct, are closely aligned (and not in conflict) with those of another organization; joining forces may increase the opportunities for each to achieve what each organization separately seeks under a broadly aligned shared agenda (NRC, 2009). Stibbe, Reid, and Gilbert (2019) argue that recognition of a collaborative advantage must also be coupled with each partnering organizations' recognition of individual value in the partnership, either through a direct strategic impact on the outcomes the organization is vested in ("mission value") or in an enhanced organizational ability to deliver its mission ("organizational value"). This concept is discussed in further detail below.

Schäferhoff, Campe, and Kaan (2009) argue that a recognition of overlapping interests is a fundamental condition for transnational partnership formation. "Norm entrepreneurs"—actors skilled at promoting and structuring the normative foundations for partnerships, persuading others to join in their efforts—can play instrumental roles in partnerships in which social learning and shared values are developed. Nevertheless, while interests may overlap, asymmetries in access and control of information, material resources, and finance, among others, can also create initial conditions of partnerships that may lead to the perpetuation of inequities in partnership activities (Contu and Girei, 2014). Some scholars have cautioned that partnerships can be formed as a result of a powerful actor mobilizing relationships largely for its benefit in terms of enhanced legitimacy, recognition, or control (Contu and Girei, 2014). In such a case, the mission of and organizational value to any one actor in a partnership would overshadow the collaborative advantage, undermining the partnership's longer-term success.

Presently, the United Nations is working to accelerate the formation of partnerships to advance the SDGs through the 2030 Agenda Partnership Accelerator.[4] This effort provides training support and advisory service by building partnership skills and competencies, including those needed to develop and implement partnerships, as well as supporting the development of policies, strategies, systems, processes, legal agreements, and culture that support collaboration. This U.N. partnership initiative responded quickly to the COVID-19 pandemic by launching two publications: an *SDG Partnership Guidebook* (Stibbe and Prescott, 2020) and a research report that compiled learning from good practices (Prescott and Stibbe, 2020).

CHARACTERISTICS OF PARTNERSHIPS FOR SUSTAINABILITY

Trust

Sustainability of partnerships are fundamentally determined by trust and shaped by the continuation of trusted relationships among people. The literature suggests that rather than solely relying on external motivators for individual compliance (e.g., punishments and rewards), it is preferable to focus on internal motivators, including trust in others (Hamm et al., 2013; Stern and Coleman, 2015; Song et al., 2019). Stern and Coleman (2015) characterize four types of trust in the context of analyzing collaborative natural resource management: (1) rational trust, based on a calculative assessment of expected benefits and risks informed by the history of performance and predictability; (2) procedural trust, which is about fairness and integrity of the procedures involved; (3) affinitive trust, which is shaped by emotions, charisma, shared identities or feelings, but not always longer-term interactions; and (4) dispositional trust, a relatively stable personality trait signaling one's predisposition to trust another entity. These four types highlight the need to take a multidimensional approach when trying to understand the role of trust in collaborative arrangements. Song et al. (2019) conclude that rational trust, which pertains to calculated risks and expectations of participation, performance, and utility, strongly predicts goal consensus. Procedural trust based on process-based notions such as integrity, fairness, and perception of equity, justice, and dignity, can partially compensate for a lack of informal interactions. Song et al. (2019) also found that affinitive trust—informal and characteristic-based aspects

---

[4] More information is available at: https://sustainabledevelopment.un.org/Partnership Accelerator.

of longer-term relationships, such as familiarity, respect, and shared experiences—were least prevalent in analyses but most significant for influencing decision making in binational resource management.

Participation

Participation is a core component of any effort of building and sustaining partnerships. The nature of a participatory process, including the conferring of respect on all sides and the chosen forms of engagement, strongly influences the structure and sustainability of collaboration. Five participation types are presented in Table 2-1, following the typology of Margerum (2008) and van Buuren, van Meerkerk, and Tortajada (2019). These different types of participation can evolve into other hybrids. Action-oriented initiatives for specific goods or services include situations in which members of the general public use a "public space" to reach their goals (van Buuren et al., 2019). Another type of participation intended to support specific

TABLE 2-1 Different Types of Invited and Created Participation

	Invited Participation			Created Participation	
Type of Participation	Capacity-driven participation	Legitimacy-driven participation	Project-oriented initiatives	Action-oriented initiatives	Policy-oriented initiatives
Description	Stakeholders are invited to participate to strengthen governance capacity	Stakeholders are invited to participate to enhance legitimacy	Stakeholders/citizens mobilize to develop their own project proposal, challenging governmental decision making affecting their interests	Stakeholders/citizens mobilize to organize and manage on-the-ground action in managing water resources (e.g. monitoring, education, restoration)	Stakeholders/citizens mobilize to change existing rules or initiate new rules and regulations for managing water resources
Motive	Empowering stakeholders is a way to enable action	Participation is a way to ensure support for policy action	To prevent public authorities from realizing their own proposal, by developing a credible alternative	To realize an initiative that adds public value	To start a policy-oriented lobby (because current policies or rules disadvantage stakeholders' position)

SOURCE: Reprinted (courtesy of Creative Commons license) from van Buuren, van Meerkerk, and Tortajada (2019).

values or rights (e.g., water access or Indigenous cultural resources) involves grassroots actions and environmental activism, by such means as agenda-setting or policy lobbying (Mazzoni and Cicognani, 2013; van Buuren et al., 2019). In many of the cases documented in the literature, a partnership or collaboration arises in the course of invited or created spaces of participation (GAO, 2008; Margerum, 2008; van Buuren et al., 2019). In practice, types of participation will vary and be adapted throughout a partnership's development, as goals evolve, learning takes place, and novel alliances are formed.

Coproduction of Knowledge

Collaborative relationships among public, private, and civil society are more productive and sustainable if they provide incentives and value to all stakeholders, rather than the ratification of one group as "the" source of knowledge or innovation over others (Contu and Girei, 2014; National Research Council [NRC], 2011). Collaboratively producing knowledge among participants in a partnership is thus fundamental to the aim of collective value creation in sustainability partnerships. Coproduction captures the idea of continual interaction between knowledge-making and decision making in the context of planning and implementation for sustainable development. Multi-stakeholder partnerships targeting sustainable development in the U.S.–Mexico border region confront complex cross-border socio-ecological system (SES) dynamics, that require both the ability to adapt and transform in response to a range of economic, cultural, political, and environmental challenges. Given that most sustainable development issues are the result of such dynamics, a diversity of knowledge is needed to effectively contribute to sustainable development (Clark et al., 2016b). In addition, partnerships are more likely to be effective when they account for the salience, credibility, and legitimacy of their knowledge production activities with the stakeholders to which they are beholden (Cash et al., 2003, p. 8086).

In regions such as the U.S.–Mexico drylands, attention to knowledge diversity may mean intentional inclusion of Native communities, as well as attention to multicultural civil society actors and public-sector actors, at all levels of administration on both sides of the border. There are significant power dynamics within coproduction processes, and these may be particularly exposed in intercultural and transnational contexts. Only recently has the coproduction of knowledge research acknowledged the need to include different cultures, languages, world views, identities, practices, and ethics in a context of asymmetries of power and rights by connecting with Indigenous and other knowledge systems (Johnson et al., 2016; Tengö et al., 2017). Co-production processes can empower some actors or some forms of knowledge more than others; partnerships can aim to address such potential asymmetries in co-production activities (Turnhout et al., 2020, p. 17). As detailed in Chapter 3, in the transnational U.S.–Mexico context, the diversity of actors and knowledge systems[5] poses a challenge to effective partnerships; coproduction implies a negotiation of shared risks and responsibilities that must be transparent to all participants in a partnership.

Tengö and colleagues (2017) argue that effective processes for knowledge co-production among partnership participants should engage in five tasks: to mobilize, translate, negotiate, synthesize, and apply multiple forms of evidence, while respecting the integrity of each knowledge system. Active commitments by knowledge holders as well as their organizations are crucial, as are processes built to increase trust and communication while accounting for language, culture, worldviews, and varying experiences. As illustrated by partnership efforts involving Indigenous communities in the U.S.–Mexico region, described in Chapter 3, these partnerships require significant time and resources. Moreover, issues of diversity, identification, representation, delegation, and liaison need to be recognized within knowledge systems (Tengö et al., 2017).

Effective partnerships will also include numerous institutional mechanisms for communication, translation, and mediation of knowledge across boundaries (Cash et al., 2003). Boundary work is a term often used to describe organizations that mediate the science-policy interface (Clark et al., 2016a) but can also apply to other forms of partnerships that do not involve the research community. Boundary work is thought to be more effective if it involves meaningful participation by stakeholders, efforts to ensure accountability to stakeholders, and the production of "boundary objects"—reports, models, maps, standards, etc.—that integrate the diversity of viewpoints within the partnership and the communities they wish to serve or influence (Clark et al., 2016a).

Alignment

Alignment has been defined as identifying synergies in order "to increase coherence, efficiency, and effectiveness for improved outcomes" among partners (Dazé et al., 2018, p. 3). Alignment of the partners' perspectives, values, and processes requires the flexibility to coordinate and integrate new information and knowledge. Three categories illustrate specific alignment characteristics: (1) informal alignment, where information is shared throughout different policy processes and collaboration

---

[5] "Knowledge systems are made up of agents, practices, and institutions that organize the production, transfer and use of knowledge" (Cornell et al., 2013, p. 61).

in implementation is ad-hoc; (2) strategic alignment, where coordination mechanisms are formally established and some joint initiatives are implemented; and (3) systematic alignment, where a shared vision for resilience, incentives for coordination across actors and levels, and implementation strategies are harmonized (Dazé et al., 2018; OECD, 2019).

Leadership

Sustainability partnerships require leaders with exceptional skills to navigate collaboration and governance approaches across diverse social, political, and cultural boundaries, targeting both sustainable development and the resilience of a complex cross-border socio-ecological region (Perz, 2019a). The co-creation of public value through partnerships requires coordination across sectors, scales, and jurisdictions (Garrick et al., 2018), a challenge that is particularly salient in international, cross-border contexts. Leaders of effective multi-stakeholder partnerships thus confer skills and knowledge on effective collaboration and boundary-crossing: that is, they share knowledge beyond addressing conventional complex environmental problems (e.g., employing biophysical and socio-economic disciplines) by tapping expertise in applied behavioral sciences. In particular, they tap into the knowledge of organizational behavior, addressing organizational culture, team dynamics and productivities, and inter-organizational relationships, including the disciplines of psychology and management (Hersey et al., 2013; Ott et al., 2008; Perz, 2019b) and the science of team formation and related social processes (Fiori, 2008; Wildman and Bedwell, 2013).

Collaborative groups need to identify trustworthy leadership, skilled at articulating the group's vision while understanding multiple sides of an issue, ensuring that the collaborative process is followed, and championing the agenda (GAO, 2008, p. 22). Furthermore, building leadership skills will allow collaborative group members to successfully represent their organizational interests (Cumiskey et al., 2019; GAO, 2008; Pulwarty and Maia, 2015; Raadgever et al., 2008; Westley et al., 2013).

Leaders of effective multi-stakeholder partnerships are charged with fostering "collaborative advantage": demonstrating that the added values expected from partnership activity can only be reached through collaborative work. Leadership aims at good co-adaptive collaborative practices for two goals, co-generating useful knowledge (Cash et al., 2017) and fostering novel innovation solutions (Kofinas et al., 2007). Achieving these goals entails the slow process of social learning (including feedback), where groups of people with shared interests proactively learn through partnership activities (Kilpatrick et al., 2003).

Choosing the most effective leadership and collaboration strategies depends largely on the developmental stage of a partnership and the internal or external challenges it is facing. Greenleaf (2002) advocates "servant leadership," when leaders serve the partners by pursuing the shared interests of the partnership rather than those of individuals, particular sectors, or actor groups. In contrast, Spillane (2006) recommends "distributed leadership;" responsibilities are to be divided among subgroups and organizations and when facing a crisis or problems, decisions are taken jointly in the light of the shared partnership goals. This distributed leadership is suitable both for partnerships with clear vertical (top-down command controlled) and those horizontal (network) collaborative structures. According to Perz (2019a), both leadership types support productive collaboration in that the former contributes to the efficient completion of work and the latter promotes the enhanced flow of information among members, thus facilitating innovation.

In a cross-border region, characterized by change and uncertainty, a leader's role in building the capacity of partnerships to collaborate on challenging issues is fundamental for sustainable development (Armitage et al., 2008; Bouwen and Tailliey, 2004). In particular, cross-border partnerships greatly enhance their effectiveness from clearly defined collaborative structures, including clarity in the functional roles of partnership constituents.

BEST PRACTICES OF SUCCESSFUL PARTNERSHIPS

The literature on partnerships suggests a mix of practices, mechanisms, and processes are being used to guide participation, co-production, and alignment for collaboration, achieving value co-creation. Beginning by characterizing and securing the common interest among multiple participants—as opposed to beginning by defining the specific practice of ecosystem, watershed, or other integrated management—allows potential partners to identify bases for partnerships and for stressing the importance of governance in realizing such collaboration (Iott, 2010). Several characteristics are especially noteworthy:

- *Inclusive representation:* Documentation on effective partnerships underscores the importance of having stakeholder participation and representation from individuals and organizations with process or outcome interests (GAO, 2008). Inclusion criteria for stakeholders may range from those necessary for implementation to those who may be impacted by possible agreements or outcomes, and including otherwise neglected groups in decisions (Brunner, 2010; GAO, 2008; Westley et al., 2013).

- *A collaborative process:* A collaborative "fit-for-purpose" process should be designed by the participants (GAO, 2008; Hazelwood, 2015; Iott, 2010). A collaborative process that utilizes a neutral facilitator with collaborative process expertise may be useful in some cases (GAO, 2008). In transcultural or transborder processes, managing cultural and language differences can be fundamental. Given the reality of asymmetrical capacities and positions, the negotiation of specific mechanisms for addressing these differences is important for building trust (Brunner, 2010; GAO, 2008; Pattberg and Widerberg, 2014, 2016).
- *Development and agreed-upon understanding of a common goal:* Partnerships should have clear goals that align with the norms and practices of participating entities. "In a collaborative process, the participants may not have the same overall interests—in fact, they may have conflicting interests" (GAO, 2008, p. 22). A fundamental premise of conflict resolution is agreement on shared facts (McCreary et al., 2007). Developing common goals and securing the common good thus require learning, trust, and time (Brunner, 2010; Pattberg and Widerberg, 2016; Schäfferhoff et al., 2009).
- *Processes for obtaining information:* "Effective collaborative processes incorporate high-quality information, including both scientific information and local knowledge, accessible to and understandable by all participants" (GAO, 2008, p. 23). Establishing processes for acquiring information is thus critical, particularly in transboundary contexts where information access is differentiated among participating entities, and different institutional norms govern information access (Garrick et al., 2018; Pulwarty and Maia, 2015).
- *Mechanisms for data and knowledge sharing:* Transparency and adequate sharing of knowledge is important for establishing trust and forming a common basis for pursuing shared goals. Data sharing can be a challenge in transnational partnerships or partnerships involving a mixture of private, public, and civil society actors with different sets of knowledge, experience, and information access. Respecting the norms and institutional constraints of participants in data sharing, while working to enhance transparency and accountability through partnership-specific data-management protocols, can thus be critical (Garrick et al., 2018; Pulwarty and Maia, 2015).
- *Leverage for available resources:* Collaboration can take time and resources to accomplish such activities as building trust among the participants, setting up the ground rules for the process, attending meetings, conducting project work, and monitoring and evaluating the results of work performed (GAO, 2008, p. 23). Taking stock and

mobilizing existing resources is important, particularly in contexts of asymmetric resource access, which characterize many transnational partnerships. Language barriers and institutional capacities also must be considered in resource allocations. Access to a diversity of financial resources can help sustain effective partnerships (Cumiskey et al., 2019; GAO, 2008; Iott, 2010; Westley et al., 2013).

- *Incentives for collaboration:* Economic and shared-value incentives can facilitate reaching goals and reduce inherent transaction costs in partnerships, recognizing the differential needs and motivations of partnering organizations (GAO, 2008). Partnerships can develop institutional arrangements that facilitate the pursuit of a common agenda, while also aiming for flexibility and adaptability to specific partner needs (Cumiskey et al., 2019; Iott, 2010; Raadgever et al., 2008; Westley et al., 2013).
- *Monitoring results for accountability:* To be effective, the participants in partnerships need to be accountable to their constituencies and to the process that they have established. Each partnering organization or entity will have specific constituencies and interests. In addition, organizations supporting the process expect accountability for the time, effort, money, or patience they invested in a partnership. Ensuring that all partners are vested in the common goal and can see mutual benefits from the partnering activities enhances accountability to the partnership as a whole. Mechanisms of accountability can involve graduated sanctions for rule violators and accessible means of dispute resolution. Accountability ideally is evaluated both internally, in relation to the partner organizations and member activities, as well as externally, in terms of the influence on sustainability outcomes (GAO, 2008; Ostrom, 1990).

SUSTAINABILITY PARTNERSHIP PERSISTENCE

For partnerships to be successful and persist over time they need clearly defined goals, roles, and responsibilities. However, effectively putting goals in place is about deciding not only on the end product, but also whether goals are created in a collaborative process (Pattberg and Widerberg, 2016). For partnerships to succeed, it is essential to engage with not only powerful and influential members but also relatively less powerful members, in a power-balancing environment. Third-party intervention can function to balance the joint influences of partners (Tandon and Chakrabarty, 2018). Leadership is considered an important ingredient throughout a partnership. The start of a partnership needs an entrepreneur or broker, "convener," or "orchestrator" (Abbott and Snidal, 2010; Glasbergen, 2010; Gray, 2007; Tandon and Chakrabarty, 2018).

In a study examining the coordination of multiple stakeholders, sustainability partnerships, and collaborative activities to reach mutual and organization-specific goals, an organizational design perspective was used to compare the decision-making processes of 94 partnerships (MacDonald et al., 2019). Results indicated "that collaborative decision-making has an indirect and positive impact on partnership capacity through systems that keep partners informed, coordinate partner interactions, and facilitate ongoing learning" (p. 409). Research on and the practice of multi-stakeholder partnerships supports the finding that partnership capacity depends on how the decision-making process is designed, in addition to internal mechanisms that manage and examine collaborative activities (MacDonald et al., 2019).

Recent research by van Buuren, van Meerkerk, and Tortajada (2019) makes distinctions among three sets of conditions important for sustaining partnerships: (1) participants' capabilities and characteristics; (2) effective interactions between authorities and participants; and (3) public institutions' response and receptivity and capacity for using a participatory process. These authors find that organizing collaborative participation efforts is vital to make the underlying values and benefits of involvement transparent and to incorporate feedback. Genuine dialogue and due deliberation, including defining problems and goals with all participants, are needed to achieve meaningful co-creation in participatory efforts. Finally, they find that relationships built on trust increase the value of information exchange, and shared learning can increase participant satisfaction and outcomes.

Drawing on the discussions and literature cited above on the characteristics and types of partnerships, the sustainability of partnerships has been shown to depend on whether processes for sustaining collaborative vision building are focused on securing the common good, facilitating knowledge building and utilization, facilitating network development both horizontally and vertically among key actors and with key actors, using policy entrepreneurs to create momentum and gain support, and pursuing flexibility and respect.

In a rapidly changing environment characterized by trends of increasing aridity, water use, and land use, and by economic and population growth on the U.S.–Mexico border, four key areas that sustain ongoing partnerships have emerged (Biggs et al., 2010; Brunner, 2010; Folke et al., 2005; Olsson and Galaz, 2012; Pulwarty and Maia, 2015; Raadgever and Mostert, 2005; Westley and Mintzberg, 1989; Westley et al., 2013):

- *Anticipation, preparation, and mobilization for change:* effectively taking advantage of forthcoming challenges and opportunities for change;
- *Recognizing or creating and engaging windows of opportunity:* understanding the importance of timing and entry points to

connect and mobilize resources and people; identifying champions and leaders at any level who are willing to take risks and convince others to take risks and to help provide institutional cover to innovators;
- *Identifying and communicating opportunities for "small wins" without losing sight of larger goals:* sustaining the ability and capacity to recognize (often small) projects that can build trust and confidence in the capabilities and intentions of the actors involved, and agreeing to take a "whole system" perspective and find mutually beneficial leverage points for learning and collaboration; and
- *Financing the deliberative process and maintenance, as well as the knowledge products and infrastructure:* ensuring that adequate public and private resources are accessible, public and private financial instruments (charges, prices, insurance, etc.) are utilized, and decision making and financing are managed together.

PRINCIPLES OF EFFECTIVE MULTI-STAKEHOLDER PARTNERSHIPS

Effective partnerships can increase the coherence of systems to deliver the greatest value toward achieving broader goals, in this case, SDG 17, with available resources. Prescott and Stibbe (2020) argue that the pursuit of effective partnerships for the SDGs requires a dynamic leader, strong champions, entrepreneurial management, risk-tolerant hosts, an adaptable business model, flexible support systems, strong connectivity, and investment in an enabling environment.

Brouwer et al. (2016) proposed seven principles for effective multi-stakeholder partnerships for sustainable development (see also Brouwer et al., 2018). In many regards, these principles synthesize the core themes laid out in the preceding sections and serve as a conceptual framework for understanding and assessing the sources of effective partnership.

- *Principle 1: Embrace systemic change.* Sustainable development involves highly complex processes and requires a commitment to iterative monitoring and evaluation routines, during which deviation from the target can be seen as an opportunity to learn and adjust rather than as a failure (Dietz et al., 2003). A systems approach benefits from diversity; as more perspectives and visions may offer a broader portfolio of opportunities and solutions for problems.
- *Principle 2: Transform institutions to induce desired change.* Rigid systems may benefit from having the rules of the game changed. That is, changes may be needed in the institutions that determine

the norms and ways people think and behave, often linked to traditions, cultural beliefs, mental models, among others. Partnerships may help induce transformations by helping partners critically view and evaluate existing institutions.
- *Principle 3: Work with the power to achieve equitable solutions.* Understanding the power structures and relations of a system is the basis for potentially bringing about change; multi-stakeholder partnerships may balance power inequality or use power structures to induce beneficial change.
- *Principle 4: Deal with conflict.* Conflict in multi-stakeholder partnerships is almost inevitable and can be necessary for change to occur. Identifying, accepting, and attending to conflict can strengthen partnerships and enhance their effectiveness.
- *Principle 5: Communicate effectively by listening to all partnership members.* Effective communication involves exploring underlying worldviews on issues, challenges, and opportunities, while allowing partners to clearly state their perspectives, ideas, and opinions.
- *Principle 6: Promote collaborative leadership.* This promotion enables stakeholders to work together, share responsibility, and develop the confidence to tackle difficult issues. One form of collaborative leadership is horizontal integrative leadership.
- *Principle 7: Foster participatory learning.* Multi-stakeholder partnerships enable actors and stakeholders to learn together by sharing knowledge and through collective experience. Organizing events and activities that foster talking, sharing, analyzing, decision-making, and reflecting on partnership activities stimulates interest and confidence in participatory learning and monitoring methods.

The committee decided to advance these principles while adding two other key qualities and considerations. First, actor involvement in multi-stakeholder partnerships follows an inclusive participatory approach in all aspects and phases of a partnership's life cycle. Multi-stakeholder partnerships for sustainable development operate on inclusive collaborative processes throughout, and their leadership style may adapt following the partnership's development, tasks, and effectiveness. Partnership members need to jointly agree on their roles and responsibilities. Second, strong and sustained partnerships develop through an iterative feedback process; thus, there is no single approach that leads to sustainable development. They also develop through a "collaborative" and trusted mechanism. Given the diversity of actors representing different disciplines and sectors, an iterative process may be as critical for sustainable development as knowledge production. One of the challenges of sustaining trust, especially in binational regions, is the rate of staff turnover (rotating positions) and dwindling program resources within agencies, and the increase of contractual positions filled with people who may not have the background and social capital to strengthen these processes (Song et al., 2019).

SUMMARY

The literature on multi-stakeholder partnership stresses the need for such partnerships to be plural in their composition. Their members need to be receptive to embracing, if not to embrace, alternative paradigms, traditions, and practices, and to be ready to cross those epistemic frontiers through an iterative process that traces unique paths for each partnership. The leadership of multi-stakeholder partnerships should share the above principles and be effective in keeping partners moving toward achieving their common goals—themselves jointly established through the concerted action of all the stakeholders.

While partnerships of this nature inevitably face complex realities, those striving to achieve the SDGs in areas or communities along the U.S.–Mexico border face an added level of complexity that results from the interaction of a demanding environment with an often intractable level of social, economic, cultural, and political asymmetries and contrasts across a border that is also a magnet for intense activity and traffic—both licit and illicit—of products and people.

FINDINGS AND CONCLUSIONS

FINDING 2-1: Complementary skills and capacities, and perceived collaborative advantage are critical elements for partnership emergence. Partnerships are also likely to emerge to fill perceived gaps in governance.

FINDING 2-2: Partnerships are characterized by thoughtful approaches to the nature and process of participation.

FINDING 2-3: Partnerships necessarily entail negotiations regarding knowledge coproduction, sharing, access, and dissemination. Governance of knowledge relationships is important for trust and transparency among partners.

FINDING 2-4: Leadership matters in partnerships; it is fundamental to establishing trust, focusing collective efforts, and steering partnerships toward goals.

CONCLUSION 2-1: Effective data sharing in transnational partnerships, or partnerships involving a mixture of private, public, and civil society actors with different sets of knowledge, experience, and information access requires respecting the norms and institutional constraints of participants with enhanced transparency and accountability through partnership-specific data management protocols.

CONCLUSION 2-2: Establishing informal community relationships and integrating indigenous and local knowledge are instrumental in partnerships that span administrative levels and geographic boundaries.

CONCLUSION 2-3: Knowledge co-production creates value in sustainability partnerships when it emanates from mutual or "horizontal" relationships among all the involved actors, confronting current power asymmetries with a commitment to combat inequality and exclusion.

CONCLUSION 2-4: Partnership persistence requires a systemic approach toward a shared goal. It is a function of the partners' organizational flexibility, adaptation to change, financial resources, and norms of distribution, as well as whether they maintain an environment that fosters innovation, learning, collaboration, and trust.

CONCLUSION 2-5: Alignment among partners to identify synergies for pursuing and securing the common good achieves coherent, efficient, and effective outcomes. Effective alignment requires flexibility in the partners' perspectives, values, and processes to enable coordination, identify appropriate entry points for new information integration, and achieve continuous learning.

REFERENCES

Abbott, K.W., and Snidal, D. (2010). International regulation without international government: Improving IO performance through orchestration. *The Review of International Organizations, 5*, 315–344. doi: 10.1007/s11558-010-9092-3.

Armitage, D., Marschke, M., and Plummer, R. (2008). Adaptive co-management and the paradox of learning. *Global Environmental Change, 18*, 86–98. doi: 10.1016/j.gloenvcha.2007.07.002.

Bäckstrand, K. (2008). Accountability of networked climate governance: The rise of transnational climate partnerships. *Global Environmental Politics, 8*(3), 74–102.

Biggs, R., Westley, F.R., and Carpenter, S.R. (2010). Navigating the back loop: Fostering social innovation and transformation in ecosystem management. *Ecology and Society, 15*(2), 9.

Blatter, J., and Ingram, H. (2000). States, markets and beyond: Governance of transboundary resources. *Natural Resources Journal, 40*, 439–471.

Börzel, T.A., and Risse, T. (2002). Public-private partnerships: Effective and legitimate tools of international governance? In E. Grande and L.W. Pauly (Eds.), *Complex Sovereignty: Reconstituting Political Authority in the 21st Century* (pp. 195–216). Toronto, Canada: University of Toronto Press.

Bouwen, R., and Taillieu, T. (2004). Multi-party collaboration as social learning for interdependence: Developing relational knowing for sustainable resource management. *Journal of Community and Applied Social Psychology, 14*(3), 137–153. doi: 10.1002/casp.777.

Brouwer, H., Woodhill, J., Hemmati, M., Verhoosel, K., and van Vugt, S. (2016). *The MSP Guide: How to Design and Facilitate Multi-stakeholder Partnerships*. Wageningen, The Netherlands: Wageningen University and Research (WUR) and Wageningen Centre for Development Innovation (WCDI) and Rugby, UK: Practical Action Publishing.

Brouwer, H., Hemmati, M., and Woodhill, J. (2018). Seven principles for effective and healthy multi-stakeholder partnerships. *ECDPM Great Insights Magazine, 8*(1). Winter 2018/2019.

Brown. L.D. (1983). *Managing Conflict at Organizational Interfaces*. Reading, MA: Addison-Wesley.

Brunner, R. (2010). Adaptive governance as a reform strategy. *Policy Sciences, 43*, 301–341.

Cash D.W., Clark, W.C., Alcock, F., Dickson, N.M., Eckley, N., Guston, D.H., Jäger, J., and Mitchell, R.B. (2003). Knowledge systems for sustainable development. *PNAS, 100*(14), 8086–8091. doi: 10.1073/pnas.1231332100.

Clark, W.C., Tomich, T.P., van Noordwijk, M., Gunston, D., Catacutan, D., Dickson, N.M., and McNie, E. (2016a). Boundary work for sustainable development: Natural resource management at the Consultative Group on International Agricultural Research (CGIAR). *PNAS, 113*(17), 4615–4622. doi: 10.1073/pnas.0900231108.

Clark, W.C., van Kerkhoff, L., Lebel, L., and Gallopin, G.C. (2016b). Crafting usable knowledge for sustainable development. *PNAS, 113*(17), 4570–4578. Available: www.pnas.org/cgi/doi/10.1073/pnas.1601266113.

Contu, A., and Girei, E. (2014). NGOs management and the value of 'partnerships' for equality in international development: What's in a name? *Human Relations, 67*(2), 205–232. doi: 10.1177/0018726713489999.

Cornell, S., Berkhout, F., Tuinstra, W., Ta'bara, J.D., Jäger, J., Chabay, I., de Wit, B., Langlais, R., Mills, D., Moll, P., Otto, I.M., Petersen, A., Pohl, C., and van Kerkhoff, L. (2013). Opening up knowledge systems for better responses to global environmental change. *Environmental Science and Policy, 28*, 60–70. doi: 10.1016/j.envsci.2012.11.008.

Cumiskey, L., Priest, S.J., Klijn, F., and Juntti, M. (2019). A framework to assess integration in flood risk management: Implications for governance, policy, and practice. *Ecology and Society, 24*(4), 17. doi: 10.5751/ES-11298-240417.

Dazé, A., Terton, A., and Maass, M. (2018). *Alignment to Advance Climate-Resilient Development: Overview Brief 1, Introduction to Alignment*. NAP Global Network. Available: www.napglobalnetwork.org.

Dietz, T., Ostrom, E., and Stern, P.C. (2003). The struggle to govern the commons. *Science, 302*(5652), 1907–1912.

Fiori, S.M. (2008). Interdisciplinarity as teamwork: How the science of teams can inform team science. *Small Group Research, 39*(3), 251–277. doi: 10.1177/1046496408317797.

Folke, C., Hahn, T., Olsson, P., and Norberg, J. 2005. Adaptive governance of social-ecological systems. *Annual Review of Environment and Resources, 30*, 441–473. http://dx.doi.org/10.1146/annurev.energy.30.050504.144511.

Fu, B., Wang, S., Zhang, J., Hou, Z., and Li, J. (2019). Unraveling the complexity in achieving the 17 sustainable-development goals. *National Science Review, 6*(3), 386–388. https://doi.org/10.1093/nsr/nwz038.

GAO (U.S. Government Accountability Office). (2008). *Natural Resource Management: Opportunities Exist to Enhance Federal Participation in Collaborative Efforts to Reduce Conflicts and Improve Natural Resource Conditions.* Report to the Chairman, Subcommittee on Public Lands and Forests, Committee on Energy and Natural Resources, U.S. Senate. GAO-08-262. Washington, DC.

Garrick, D.E., Schlager, E., De Stefano, L., and Villamayor-Tomas, S. (2018). Managing the cascading risks of droughts: Institutional adaptation in transboundary river basins. *Earth's Future*, 6, 809–827. Available: https://doi.org/10.1002/2018EF000823.

Glasbergen, P. (2010). Global action networks: Agents for collective action. *Global Environmental Change*, 20(1), 130–141.

Gray, B. (2007). The process of partnership construction: Anticipating obstacles and enhancing the likelihood of successful partnerships for sustainable development. In P. Glasbergen, F. Biermann, and A.P.J. Mol (Eds.), *Partnerships, Governance and Sustainable Development*: *Reflections on Theory and Practice* (pp. 29–48). Northampton, MA: Edward Elgar Publishing, Inc.

Greenleaf, R.K. (2002). *Servant Leadership: A Journey into the Nature of Legitimate Power and Greatness.* 25th-anniversary edition. Mahwah, NJ: Paulist Press.

Gurzawska, A. (2020). Towards responsible and sustainable supply chains – Innovation, multi-stakeholder approach and governance. *Philosophy of Management, 19*, 267–295.

Hamm, J., PytlikZillig, L., Herian, M., Tomkins, A., Dietrich, H., and Michaels, S. (2013). Trust and intention to comply with a water allocation decision: the moderating roles of knowledge and consistency. *Ecology and Society*, 18(4).

Hazlewood, P. (2015). *Global Multi-Stakeholder Partnerships: Scaling Up Public-Private Collective Impact for the SDGs. Independent Research Forum 2015.* Washington, DC: World Resources Institute.

Hersey, P.H., Banchard, K.H., and Johnson, D.E. (2013). *Management of Organizational Behavior: Leading Human Resources, 10th edition*, Boston, MA: Pearson.

Ingram, H., Milich, L., and Varady, R. (1994). Managing transboundary resources: Lessons from Ambos Nogales. *Environment, 36*(4), 6–38. doi: 10.1080/00139157.1994.9929996.

Iott, S. (2010). Policy sciences and congressional research: Making sense of the research question. *Policy Sciences, 43*, 289–300.

Johnson, J.T., Howitt, R., Cajete, G., Berkes, F., Louis, R.P., and Kliskey, A. (2016). Weaving Indigenous and sustainability sciences to diversify our methods. *Sustainable Science, 11*, 1–11.

Kilpatrick, S., Barrett, M., and Jones, T. (2003). Defining Learning Communities. Paper presented at the Joint AARE/NZARE Conference, Auckland. Available: https://www.aare.edu.au/data/publications/2003/jon03441.pdf.

Kofinas, G.P., Herman, S.J., and Meek, C. (2007). Novel problems require novel solutions: Innovation as an outcome of adaptive co-management. In D. Armitage, F. Berkes, and N. Doubleday (Eds.), *Adaptive Co-Management: Collaboration, Learning and Multi-scale Governance* (pp. 249–267). Vancouver, BC: UBC Press.

Long, F.J., and Arnold, M.B. (1995). *The Power of Environmental Partnerships.* Dryden Press Series in Management. Hinsdale, IL: Dryden Press.

MacDonald, A., Clarke, A., and Huang, L. (2019). Multi-stakeholder partnerships for sustainability: Designing decision-making processes for partnership capacity. *Journal of Business Ethics, 160*, 409–426. doi: 10.1007/s10551-018-3885-3.

Margerum, R.D. (2008). A typology of collaboration efforts in environmental management. *Environmental Management, 41*(4), 487–500. doi:10.1007/s00267-008-9067-9.

Mazzoni, D., and Cicognani, E. (2013). Water as a commons: An exploratory study on the motives for collective action among Italian water movement activists. *Journal of Community and Applied Social Psychology, 23*(4), 314–330. doi:10.1002/casp.2123.

McCreary, S.T., Gamman, J.K., Brooks, B. (2007). Refining and testing joint fact-finding for environmental dispute resolution: Ten years of success. *Mediation Quarterly, 18*(4), 329–348.

Meadow, A.M., Ferguson, D.B., Guido, Z., Horangic, A., Owen, G., and Wall, T. (2015). Moving toward the deliberate coproduction of climate science knowledge. *Weather, Climate, Society, 7*(2), 179–191. doi: 10.1175/WCAS-D-14-00050.1.

NRC (National Research Council). (2009). *Enhancing the Effectiveness of Sustainability Partnerships: Summary of a Workshop*. Washington, DC: The National Academies Press. doi: 10.17226/12541.

———. (2011). *Building Community Disaster Resilience Through Private-Public Collaboration*. Washington, DC: The National Academies Press. doi.org/10.17226/13028.

OECD (Organisation for Economic Co-operation and Development). (2019). Aligning Development. (2019). *Aligning Development Co-operation and Climate Action: The Only Way Forward*. Paris: The Development Dimension, OECD Publishing.

Olsson, P., and Galaz, V. (2012). Social-ecological innovation and transformation. In A. Nicholls and A. Murdock (Eds.), *Social Innovation: Blurring Boundaries to Reconfigure Markets* (pp. 223–243). Basingstoke, UK: Palgrave Macmillan.

Ostrom, E. (1990). *Governing the Commons: The Evolution of Institutions for Collective Action*. Cambridge, England: Cambridge University Press.

Ott, J.S., Parkes, S.J., and Simpson, R.B. (2008). *Classic Readings in Organizational Behavior*. 4th edition, Belmont, CA: Wadsworth.

Pattberg, P., and Widerberg, O. (2014). *Transnational Multi-Stakeholder Partnerships for Sustainable Development: Building Blocks for Success*. Amsterdam: IVM Institute for Environmental Studies.

———. (2016). Transnational multi-stakeholder partnerships for sustainable development: Conditions for success. *Ambio, 45*(1), 42–51. doi: 10.1007/s13280-015-0684-2.

Perz, S.G. (2019a). Crossing boundaries for collaboration in comparative perspective: Key insights, lessons learned, and recommendations for future practice. In S.G. Perz (Ed.), *Collaboration Across Boundaries for Socio-Ecological System Science: Experiences Around the World* (pp. 395–429). Cham, Switzerland: Springer.

———. (2019b). Introduction: Collaboration across boundaries for socio-ecological systems science. In S.G. Perz (Ed.), *Collaboration Across Boundaries for Socio-Ecological System Science: Experiences Around the World* (pp. 1–33). Cham, Switzerland: Springer. Available: https://doi.org/10.1007/978-3-030-13827-1.

Peterson, K., Mahmud, A., Bhavaraju, N., and Mihaly, A. (2015). *The Promise of Partnerships: A Dialogue between INGOs and Donors*. New York: Farrar, Straus, and Giroux.

Pinkse, J., and Kolk, A. (2012). Addressing the climate change—Sustainable development nexus: The role of multistakeholder partnerships. *Business and Society, 51*(1), 176–210.

Prescott, D., and Stibbe, D. (2020). *Partnership Platforms for the SDGs: Learning from Practice. The Partnering Initiative and UNDESA*. Available: https://sustainabledevelopment.un.org/content/documents/2699Platforms_for_Partnership_Report_v0.92.pdf.

Pulwarty, R.S., and Maia, R. (2015). Adaptation challenges in complex rivers around the world: The Guadiana and the Colorado Basins. *Water Resources Management, 29*, 273–293.

Raadgever, G.T., and Mostert, E. (2005). Transboundary River Basin Management: State-of-the-art review on transboundary regimes and information management in the context of adaptive management, NeWater Report Series, Deliverable 1.3.1. TU Delft: RBA Centre.

Raadgever, G.T., Mostert, E., Kranz, N., Interwies, E., and Timmerman, J.G. (2008). Assessing management regimes in transboundary river basins: Do they support adaptive management? *Ecology and Society, 13*(1), 14.

Schäferhoff, M., Campe, S., and Kaan, C. (2009). Transnational public-private partnerships in international relations: Making sense of concepts, research frameworks, and results. *International Studies Review, 11*(3), 451–474.

Song, A., Temby, O., Kim, D., Cisneros, A., and Hickey, G.M. (2019). Measuring, mapping and quantifying the effects of trust and informal communication on transboundary collaboration in the Great Lakes fisheries policy network. *Global Environmental Change*, 4, 6–18.

Spillane, J.P. (2006). *Distributed Leadership*. San Francisco, CA: Josey-Bass.

Stern, M.J., Coleman, K.J. (2015). The multidimensionality of trust: Applications in collaborative natural resource management. *Society and Natural Resources*, 28(2), 117–132.

Stibbe, D., and Prescott, D. (2020a). *The SDG Partnership Guidebook: A Practical Guide to Building High Impact Multi-Stakeholder Partnerships for the Sustainable Development Goals*. The Partnering Initiative, United Nations. Available: https://sustainabledevelopment.un.org/content/documents/26627SDG_Partnership_Guidebook_0.95_web.pdf.

Stibbe, D.T., Reid, S., and Gilbert, J. (2019). *Maximising the Impact of Partnerships for the SDGs: A Practical Guide to Partnership Value Creation*. The Partnering Initiative, United Nations. Available: https://sustainabledevelopment.un.org/content/documents/2564Maximising_the_impact_of_partnerships_for_the_SDGs.pdf.

Sverdrup-Jensen, S., and Nielsen, J. (1998). *Co-management in Small-Scale Fisheries: A Synthesis of Southern and West African Experiences*. Available: https://www.oceandocs.org/handle/1834/617.

Tandon, R., and Chakrabarty, K. (2018). Partnering with higher education institutions for SDG 17: The role of higher education in multi-stakeholder partnerships. In *Approaches to SDG17 Partnerships for Sustainable Development Goals (SDGs)*. Barcelona: Global University Network for Innovation. Available: www.guninetwork.org.

Tengö, M., Hill, R., Malmer, P., Raymond, C.M., Spierenburg, M., Danielsen, F., Elmqvist, T., and Folke, C. (2017). Weaving knowledge systems in IPBES, CBD and beyond—Lessons learned for sustainability. *Current Opinion in Environmental Sustainability*, 26–27, 17–25. doi: 10.1016/j.cosust.2016.12.005.

Turnhout, E., Metze, T., Wyborn, C., Klenk, N., and Louder, E. (2020). The politics of co-production: Participation, power, and transformation. *Current Opinions in Environmental Sustainability*, 42(February), 15–21. doi: 10.1016/j.cosust.2019.11.009.

United Nations. (2015). *Transforming Our World: The 2030 Agenda for Sustainable Development*. Available: https://www.un.org/ga/search/view_doc.asp?symbol=A/RES/70/1&Lang=E.

_____. (2020). *The Sustainable Development Goals Report: 2020*. Available: https://unstats.un.org/sdgs/report/2020/The-Sustainable-Development-Goals-Report-2020.pdf.

van Buuren, A., van Meerkerk, I., and Tortajada, C. (2019). Understanding emergent participation practices in water governance. *International Journal of Water Resources Development*, 35(3), 367–382. doi: 10.1080/07900627.2019.1585764.

van Huijstee, M., and Glasbergen, P. (2010). Business-NGO interactions in a multi-stakeholder context. *Business and Society Review*, 115(3), 249–284.

Westley, F., and Mintzberg, H. (1989). Visionary leadership and strategic management. *Strategic Management Journal*, 10(S1), 17–32. Available: http://dx.doi.org/10.1002/smj.4250100704.

Westley, F.R., Tjornbo, O., Schultz, L., Olsson, P., Folke, C., Crona, B., and Bodin, Ö. (2013). A theory of transformative agency in linked social-ecological systems. *Ecology and Society*, 18(3), 27. doi: 10.5751/ES-05072-180327.

Wildman, J.L., and Bedwell, W.L. (2013). Practicing what we preach: Teaching teams using validated team science. *Small Group Research*, 44(4), 381–394. doi: 10.1177/1046496413486938.

3

Opportunities and Challenges for U.S.–Mexico Sustainability Partnerships

Chapter 2 discussed the characteristics of sustainability partnerships and how they can be sustained. Successful partnerships emerge from the entrepreneurial activities of conveners with big aspirations and a strong commitment to challenge the status quo. This chapter focuses on what makes successful partnerships effective in the U.S.–Mexico border region by summarizing discussions and insights gathered at a virtual seminar, "Sustainability Partnerships in the U.S.–Mexico Drylands Region," held in July 2020.[1] By hearing from stakeholders in the region, this activity served as the primary source of information gathering for committee deliberations.

To plan the webinar, the committee developed a bilingual online questionnaire to generate a list of potential speakers and attendees; see Chapter 1 for an explanation of the process. The selected panelists had varying tenures and areas of expertise: see Appendix C for the webinar agenda with participant's names and institutional affiliations. The panel participants discussed their work, while also underscoring sources of effectiveness in one or more of their partnership experiences.[2] Because the widespread and profound COVID-19 effects led to cross-border travel restrictions, intermittent closure of the border, and, ultimately, the shift of the planned seminar from

---

[1] The webinar can be viewed at: https://www.nationalacademies.org/event/07-15-2020/sustainability-partnerships-in-the-us-mexico-drylands-region-a-binational-consensus-study-virtual-public-seminar.

[2] Unless cited as a direct quote by a participant or as an external source, the points noted throughout this chapter, except for the committee's conclusions at the end, represent a compilation of the webinar discussions.

an in-person to a virtual meeting, the shocks and stressors of the pandemic form a backdrop to this overview of current partnerships.

The committee recognized that the questionnaire responses and panel discussions did not constitute a consistent or exhaustive dataset about the status of U.S.–Mexico partnerships and so did not set out to evaluate the partnerships discussed at the webinar. However, the stakeholders' feedback on forming and sustaining partnerships, improving communication among partners, and planning for present and future uncertainties in the operating environment served as the cornerstone for the committee's work.

HOW BINATIONAL PARTNERSHIPS EMERGE AND EVOLVE

Binational problems require binational solutions
—Irasema Coronado (Arizona State University)

Stakeholders operating in the region have found that state and national policies are usually at odds with border needs and priorities, and partnerships originating at the border aim to engage local actors in ways that go beyond the conventional state-led, top-down approach. Yoselín Cárdenas (Consejo Empresarial Nogales, A.C.) believes that local citizens and local activists are most aware of an area's economic, cultural, and environmental challenges. Andy Carey (Border Philanthropy Partnership [BPP]) said that his organization emphasizes this type of holistic, multi-directional collaboration—bringing members of government, business, academia, and nonprofit organizations to the table to share their commitment, expertise, and knowledge.

Co-Creation and Capacity Building

Turning participation into co-creation is challenging in the binational region due to the asymmetries, power dynamics, and knowledge systems, but partners find creative ways to tackle these challenges. Binationally, there is asymmetry and imbalance across several sectors regarding access to resources. When the North American Free Trade Agreement (NAFTA) was in effect, many Mexican nongovernmental organizations (NGOs) were funded by U.S. NGOs. Coronado noted that the agenda at the U.S.–Mexico border is often driven by the stakeholder with the greatest resources.

The webinar panelists supported the premise that capacity building between the United States and Mexico reduces resource and knowledge imbalances, allowing initiatives to arise from both countries. The El Paso Community Foundation (Texas) and Fundación Paso del Norte para Salud y Bienestar (Ciudad Juárez) are one example of sister organizations that emerged on either side of the border to address shared challenges.

Other notable examples of organizations working to build transborder capacity are the Fundación del Empresariado Chihuahuense in Chihuahua and the Desarrollo Económico in Ciudad Juárez. Other organizations, like BPP, engage NGOs to strengthen their capabilities to address issues of prosperity, equity, and opportunity along both sides of the border. Zach Hernandez (San Diego Association of Governments [SANDAG]) said that maintaining robust communication is central to his organization's binational partnership strategy. SANDAG has institutionalized its communication framework in a way that enables the Mexican government to formally participate in regional planning. This approach has aided in processes such as binational transportation planning.

A webinar participant asked panelists whether they had seen an asymmetry in the initiation of binational partnerships—whether more were started in Mexico than in the United States, or vice versa—and if there was such an imbalance, why they believed that to be the case. The ensuing discussion focused on the fact that in many instances, U.S. NGOs have more resources than their Mexican counterparts. An example is the Colorado River Delta Water Trust, a robust and effective partnership to restore the Colorado River Delta. Pronatura Noroeste, a Mexican NGO, is aided by the U.S.-based Sonoran Institute and the Environmental Defense Fund to establish mechanisms to acquire water from Mexican farmers at market value to restore riparian habitats, adjacent forests, and ecologically and economically important wetlands in the delta.

Gabriela Múñoz-Meléndez (El Colegio de la Frontera Norte) pointed out that NGOs and universities face challenges, or impose their restrictions, in receiving, distributing, and administering funds. Universities often take 20–50 percent of the funds as overhead. These examples are set in the backdrop of other, related asymmetries, notably the differences in water management regimes between the two countries. Despite the presence and largely effective work of a binational water institution, the International Boundary and Water Commission in the United States and its Mexican counterpart, la Comisión Internacional de Límites y Aguas, the two countries have very different federal, state, and local legal and institutional arrangements for water resource management and policy; see discussion in Appendix D.

INDIGENOUS COMMUNITY PARTNERSHIPS ACROSS THE U.S.–MEXICO BORDER

Indigenous tribes and communities have their territories and boundaries that often predate and do not align with the border established between the United States and Mexico. They also have their leaders who operate within an autonomous, independent government. Blake Gentry (Líderes Tradicionales de O'odham en México) mentioned that while Indigenous tribes are federally recognized in the United States, there is no parallel process in Mexico—a difference that may create challenges in binational partnerships with these communities. Speakers of the O'odham language are spread across 1,200 miles on both sides of the international border. Gentry said his organization works with the O'odham peoples in Sonora, Mexico, and with the Mexican government, advocating for O'odham's Indigenous rights and longstanding cultural traditions. Gentry believes that neither the United States nor the Mexican state and federal governments has prioritized tribes and that the Mexican government often ignores and represses the O'odham.

Historically, there has consistently been a census undercount of the O'odham population, which Gentry noted is problematic because public funding is contingent on population size. Since 1959, the Mexican government has officially counted only 300 O'odham in Sonora, and it annually assigns a budget for indigenous O'odham based on that number. Since this is a tribal nation without a centralized government, this apparent undercount has resulted in the O'odham being continually underserved. Because of this history, O'odham in Sonora now take their own annual census. The most recent estimated population in this census was 7,000–8,000. Assisting O'odham Indigenous community leaders, college preparatory students from the United World College in Las Vegas, New Mexico, and college students of the Border Studies Program of Earlham College in Richmond, Indiana, engaged in the launching of the census project, known as O'odham Kuinta.

The Alianza Indígena Sin Fronteras partnership, led by Octaviana Valenzuela Trujillo (Northern Arizona University), a Yaqui[3] citizen, works with Indigenous people on both sides of the border. Alianza Indígena Sin Fronteras hosts listening sessions with tribes to hear about community concerns, with an emphasis on building trust, developing long-term relationships, and creating common ground between tribes and partner agencies.

Trujillo discussed how the construction of the border wall is creating issues in places like Quitobaquito, Arizona, by depleting groundwater, dynamiting sacred terrain, causing ecological damage, and negatively affecting other resources. The wall also threatens to disrupt age-old Indigenous traditions of pilgrimage to sacred sites. Several participants asserted that the construction of the border wall is a violation of O'odham's sovereignty. Attempts to stop development on Mexican tribal lands have failed. Múñoz-Meléndez commented that Mexican international relations limit tribes' jurisdiction, which conflicts with International Labour Convention No. 169 regarding the extension of rights to Indigenous peoples and the preservation of their culture.

---

[3] The Yaqui are an Indigenous people centered in southern Sonora, Mexico.

Trevor Hare (Watershed Management Group) noted that the border wall being constructed is causing problems in what he called "our water ways," and he did not see any type of coherent strategy along the border to deal with the current circumstances. In particular, Quitobaquito, which is a sacred Indigenous site in the Oregon Pipe National Monument, is being affected: water is being used to build the wall and depleting the groundwater. In addition, Hare said, the O'odham and other Indigenous people will not be able to do what they have done since time immemorial, to go to Quitobaquito.

Gentry explained that working with Indigenous communities without understanding their protocols can do more harm than good. Indigenous communities have unique histories, cultures, and environmental needs. A webinar participant commented that a lack of outreach and extension skills on the part of most non-Indigenous organizations limits their ability to make appropriate contact with rural communities; thus, it is imperative to figure out how to ensure Indigenous voices are included in the right conversations. Panelists agreed that prior consent and prior consultation are of utmost importance. Múñoz-Meléndez added that while learning about Indigenous people's challenges and capacities can take time and effort, it is key to building and sustaining effective relationships. Trujillo underscored the importance of understanding tribal resolutions in the areas of interest, and of listening to tribal councils about the interconnections between religion and biodiversity issues. She added that tribal councils on both sides of the border can advise partnership on the needs of their constituents. (For more general discussion on tribal nations and Indigenous communities along the U.S.–Mexico border, see Appendix D.)

HOW ORGANIZATIONS CONNECT AROUND SUSTAINABILITY CHALLENGES

Government diplomacy and planning processes vary not only between the United States and Mexico but also from state to state in both countries. Historically, the countries' respective capitals, Washington, D.C., and Mexico City have controlled the binational agenda. Because decision makers in the national capitals are often disconnected from the realities of the region, subnational diplomacy at the state and city levels has become prominent in transboundary sustainability partnerships. Carey mentioned how border mayors meet regularly, as do border legislators. Border governors also meet, but not as frequently. In addition, nongovernmental institutions, organizations, and formal and informal alliances have been established to strengthen border relations.

Formal, multisectoral partnerships are key to orchestrating long-term responses to binational and bidirectional sustainability challenges, such as those concerning water and pollution. Hare observed that strong binational connections and funding sources are essential to tackling pollution in urban areas and erosion in wildlands. James Callegary (U.S. Geological Survey) described prior attempts to regulate contaminants in the Douglas, Arizona, and Agua Prieta, Sonora, border region and in Ambos Nogales (Nogales, Sonora, and Nogales, Arizona), noting that the success of the joint efforts varied in each region. Yoselin Cardenas (Consejo Empresarial Nogales, A.C.) said that in Ambos Nogales (as with other border areas), though the sustainability challenges are shared, feedback from each state's stakeholders regarding the region's priorities may differ widely, which can stymie collaborative efforts. Achieving consistency across borders is made more difficult given that political administrations in Mexico are on a 3-year cycle, while U.S. officials operate on a 2- or 4-year cycle. The outcomes of both countries' elections may also play a role in the future of many partnerships. These changes can have major impacts on formal partnerships, but informal partnerships often have the capacity to withstand them.

While fostering working relationships can set the stage for dialogue and action-oriented collaboration between partners, personal relationships are often key to bringing together organizations and growing networks. Partners have to strategically navigate the informal-formal balance to make partnerships sustainable over time. Hernandez said that SANDAG wears many hats, making connections through various channels, within and absent of a formal structure. Carey added that diplomatic channels and personal relationships are key to BPP's success.

Establishing a mix of formal and informal relationships enables partners in the binational drylands region to approach solutions from multiple angles. Benjamin Wilder (Next Generation Sonoran Desert Researchers) noted that sustainability work in the Sonoran desert consists of both top-down initiatives, such as bank- and government-funded solar plants and bottom-up initiatives, including food-water-energy nexus projects and community training conducted by academic and grassroots organizations. Agrivoltaics, a practice of co-developing solar plants and agricultural farms to boost both energy and food production, is a bottom-up example. Wilder added, bottom-up initiatives rely on bottom-up ingenuity. Bottom-up initiatives have been shown to be more responsive to local needs than top-down initiatives, which panelists agreed are often disconnected from local needs.

Partners use cross-jurisdictional and collaborative mechanisms to navigate different knowledge. The Comisión Nacional de Áreas Naturales Protegidas funded a multiyear climate study that used self-implemented designs from Indigenous peoples. The project helped people outside the region see that Indigenous communities are not homogenous; while they have shared challenges, each group takes a unique approach to address them. Wilder said that the process of bringing together researchers and government organizations to provide support for specific issues, such as health or access to freshwater, is a long process that requires listening and patience, but it leads to partnerships that stand the test of time. Establishing a long-term presence can improve consistency and trust and strengthen relationships. Panelists emphasized the need for NGOs and universities to consult with Indigenous communities in the region before embarking on projects, which aligns with the United Nations' Office of Human Rights call for free, prior, and informed consent of Indigenous peoples.[4] Múñoz-Meléndez reinforced this point by underscoring the importance of organizations taking time to build relationships with local stakeholders.

Gabriel Armenta (Índex Nogales, Asociación de Maquiladoras de Sonora, A.C.) noted that his organization, which brings together the maquiladoras (factories) in Ambos Nogales, is based in the United States and has strong relationships with financial institutions that aid the organization in initiating binational collaboration on environmental and social development matters. Involving local governments in these partnerships, however, remains challenging. The maquiladora industry participates in binational commissions and committees in various fields, including security, foreign trade, and customs. The main goals are to create jobs, improve training, improve quality of life, and foster sustainable development. Armenta noted that creating effective communication channels with the government to garner support for these projects, as well as communicating with immigration offices in local municipalities, is a critical success factor of maquiladora partnerships for sustainability. Múñoz-Meléndez observed that it is important to pay attention to the reasons why larger organizations such as government agencies and higher education institutions involve themselves in local affairs. For a binational sustainability partnership to be effective, it should understand the structure of decision making on both sides of the border. Partners rely on informal networks to keep partnerships alive, but engaging the right interlocutors is a key challenge. Múñoz-Meléndez added that partnerships need the right representations for effective change.

ROLE OF INFORMATION IN SUSTAINING PARTNERSHIPS

Building and sustaining successful relationships that are the core of partnerships requires effectively facilitating the flow of information. Carey emphasized the need to continue educating people about the region and engaging border stakeholders in various projects to help build confidence around action. Hare and Callegary noted that the opposite condition—that is, a lack of information—is often an obstacle to successful partnerships. Hernandez emphasized the importance of clear information flows among partners in helping

---

[4] More information is available at: http://www.fao.org/3/a-i6190e.pdf.

build relationships; institutionalizing input, dialogue, forum, and information sharing are crucial in building relationships. He also noted, however, that there is a learning curve that comes with working with partners in Mexico. Having Mexico represented through the voice of its binational partners in policy making is key to successful binational planning.

Academic partnerships across the border involve internships, student and professor exchanges, and in-person presentations. Many Mexican students attend school in the United States, and vice versa, especially at local public universities (though the COVID-19 pandemic may have changed the rate). Engaging the next generation is key to sustaining partnerships. Arizona State University's School of Transborder Studies works binationally to cultivate future leadership by including students in its work.

Wilder has noticed a generation of new researchers focused on the binational region and multinational, cross-discipline collaboration. Interest is increasing such that the number of people interested in teaching about the region at higher education institutions and conducting relevant research exceeds the number of positions currently available. He believes that the solution is to think about ways to undertake new research and to create new partnerships. Hare also noted the importance of engaging with and mentoring youth in the United States and Mexico around border sustainability issues. Trujillo noted how the Healing the Border Project organizes community hearings and helps youth create digital stories about the region.

THE EFFECTS OF COVID-19 ON SUSTAINABILITY PARTNERSHIPS

Trujillo noted how the COVID-19 pandemic has hindered the ability of the Alianza Indígena Sin Fronteras to hold in-person meetings, which slows relationship-building. She added that while Zoom meetings and conversations on WhatsApp are useful to facilitate communication when in-person contact is not possible, it is no substitute for in-person contact when it comes to developing relationships, strengthening alliances, establishing new allies, and building solidarity.

The border has become particularly important recently, as Carey noted, for the shipment of personal protective equipment for COVID-19 from the United States to Mexico. Expediting the shipment of supplies to address issues emerging around COVID-19 requires leveraging formal and informal relations. Several webinar participants noted ways in which the pandemic has negatively changed the nature of communication at the border, but also mentioned ways that the move to virtual engagement has facilitated communication.

Before the COVID-19 pandemic, there was a bureaucratization of partnerships, noted Cota de Yáñez (committee member), and a formal regularity of meetings with set agendas. Though the structure of business has changed during the pandemic, the flow of information has not stopped. Many binational partnerships, covering such activities as maquiladoras, rail transport, and medical services, continue to play critical roles. Panelists discussed how in this era, informal networks have been strengthened to an extent never seen before, aided by access to communications technology (e.g., cellphones, Whatsapp).

It is not known whether formal structures and hierarchies will reassert themselves after the COVID-19 pandemic has passed or whether partnership members will see advantages in the informal nature of communications and decision making and continue in that fashion. Commerce between the two countries continues to rely heavily on technology, and participants expressed the hope that partnerships can withstand the changes that have resulted from the pandemic through high-tech tools and information technology services for both Mexican and U.S. partners. Panelists were hopeful that the significant investments that stakeholders have made into forming relationships around sustainability are only being strengthened during the pandemic and will continue after it is over.

Before the COVID-19 pandemic, health service networks in the U.S.–Mexico border region were managed very formally and from the top down: in Mexico, these networks started with the national health secretary, and included officials at the federal, state, and municipal levels; in the United States, the networks are managed by the U.S. Department of Health and Human Services (HHS). Carey told panelists about the U.S.–Mexico Border Health Commission,[5] which began as a binational commission in 2000 between the HHS Secretary and the Mexican Secretary of Health. He commented that even though then-President Trump had somewhat sidelined the U.S. work of the commission during his administration, partnerships at the border have continued to thrive. There have been successful joint planning around infectious disease, juvenile diabetes, and cancer, led mainly by state governments. Carey noted the severe impacts of the pandemic at the border region: [at the time of the workshop,] hospitals were overrun, and there was a lack of personal protective equipment. BPP and other organizations have been working to expedite the crossing of in-kind medical supplies in the border region. He said that when there are gaps in border health service coordination, particularly at the political level, it can create significant challenges for providers.

When COVID-19 was declared a pandemic in March 2020, essential services provided by partnerships could not go under lockdown, so each partnership had to decide what worked best for it in terms of continuing collaboration. With border crossing restrictions, agency closings,

---

[5] More information is available at: https://www.hhs.gov/about/agencies/oga/about-oga/what-we-do/international-relations-division/americas/border-health-commission/index.html.

cancellation of in-person meetings, stay-at-home orders, and many stakeholders dealing with personal and family illness, networks such as the one between Sonora and Arizona started to rely on informal communication. Examples of such communications at the peer-to-peer level include drugstore owners networking directly with customers, Mexican physicians communicating directly with their colleagues in the United States, and families forming a community with other families. On the institutional level, local public health agencies relied on official updates in each country to make decisions affecting their communities. Local stakeholders met as needed, using virtual platforms, to discuss the status of private and public hospitals, both COVID and non-COVID facilities.

The webinar panelists all said that the true test of the strength of the binational partnerships will be how they fare during and after this pandemic. However, the panelists noted that the binational response to the health crisis would not have been possible if partnerships had not been formed and developed before COVID: the trust, confidence, and personal involvement that had been cultivated before the pandemic were essential to continued collaboration. Cardenas noted that although the pandemic has changed the way Consejo Empresarial Nogales communicates, many of the organization's relationships are stronger now than ever before.

While many health-related partnerships continue to thrive, others have been severely strained by the pandemic. An example is ARSOBO (an acronym for Arizona-Sonora Border), which is a binational collaboration among academia, businesses, NGOs, individual patients, the U.S. Consulate, municipal governments of the state of Sonora, and volunteers from the United States. The volunteers would come to the border area every 6 weeks or so to provide health services, such as therapies, audiometer tests, hearing-aid adaptations, and prosthetic measurements. Under COVID restrictions, students and others are not allowed by their U.S. academic institutions to cross the border, and Mexicans were also prohibited from crossing for a time. In response, a new virtual partnership has developed. Based probably on the strong personal relationships among the members on each side, subnetworks connect by specialty areas, including physical therapy (with U.S. volunteers and Mexican translators, mostly), audiology (U.S. medical doctors and Mexican technicians), and prosthetics (U.S. Hanger, Inc., a prosthetic manufacturer, and local technicians they trained in Nogales, Sonora).

All these partnership initiatives relied on monthly or more frequent visits, often back and forth daily, due to the age of many of the patients and the border closure to nonessential travel. These new partnerships have relied on Mexican staff accommodating patients' schedules to the U.S. participants' working hours (mostly from home using virtual technology), as well as finding volunteers to translate virtually. Patients from all over northern Mexico communicate via Facetime with the Mexican ARSOBO staff, who in turn connect via Zoom with University of Arizona academic staff, volunteers, board members, Hanger, Inc., and others. For follow-up, patients were provided with future Zoom appointments to reduce the number of trips and exposure at the border. Many of the patients are diabetics, children, and the elderly—all at high risk of health complications.

This 12+-year partnership would not exist today if not for the commitment of the participants. All manner of challenges have been faced, from city-to-city cross-border coordination, to the dollar-to-peso exchange rate, to the taxation of imports even for NGOs, to the perception of border violence, to border-crossing bans by universities and corporations. The U.S. Consulate played a crucial role for staff and volunteers without visas.

WHAT SUCCESSFUL PARTNERSHIPS LOOK LIKE

Successful partners challenge each other and defy the status quo.
Set BIG goals and don't settle for minimal progress.
   —Zachary Hernandez (SANDAG)

Chapter 2 discussed the characteristics of sustainability partnerships and illustrated how partnerships ensure their sustainability. Successful partnerships emerge from the entrepreneurial activities of conveners with big aspirations and a strong commitment to challenge the status quo. This section focuses on what makes successful partnerships special in the U.S.–Mexico border region.

Partnerships that succeed share mutually beneficial goals that are well-grounded in the unique characteristics of the region. The Border Health Commission's role is to "bring together the two countries and their border states to address border health challenges by providing the necessary leadership to develop coordinated and binational actions that can improve the health and quality of life of all border residents."[6]

Partnerships are bidirectional. Partners of the Americas connects higher education institutions across borders to exchange knowledge, build programs, and foster long-term partnerships.[7] BPP has built the capacities of NGOs on both sides of the border, enabling initiatives to emerge from both countries. Partnerships build relationships to maintain an enabling environment to take advantage of opportunities and collaborate across sectors, often with a sense of urgency. For example, organizations across

---

[6] More information is available at: https://www.hhs.gov/about/agencies/oga/about-oga/what-we-do/international-relations-division/americas/border-health-commission/index.html.

[7] More information is available at: https://partners.net/.

the border have shown remarkable cooperation to respond to COVID-19 with personal protective equipment.

Successful partners have a binational mindset that acknowledges that challenges on one side of the border affect the other side. The work of partners is crucial to educate leaders in both countries on the complexity of the region and develop the future leaders that successful partnerships require in government, in NGOs, among academics, among indigenous communities, and in the private sector. Universities in the region address important and complex topics in the U.S.–Mexico border region, including migration, health, and applied social policy; media and expressive culture; culture, language, and learning; and U.S. and Mexican regional immigration policy and the regional economy. El Colegio de la Frontera Norte does research and education on the complex regional social processes with a multidisciplinary perspective.

Successful partnerships rely on participatory mechanisms to achieve effective co-creation. They do so by tackling the challenges of asymmetries in the region. Alianza Indígena Sin Fronteras works with Indigenous people across the border to affirm their rights and listen to their concerns on policy decisions, such as those concerning the border and COVID-19 travel restrictions, that have an impact on the pilgrimage to the sacred site of the O'odham people. SANDAG effectively manages the complexity of subnational diplomacy to achieve participatory transportation planning that includes Mexico and builds the public trust to engage the private sector constructively.

"Success" is a relative term, and every webinar panelist discussed those factors that they deemed critical for the type of success that they hoped to achieve from the partnerships in which they are involved. In some instances, success might be defined in terms of the explicit objectives of the partnership. In other instances, success was defined in more process-based and subjective terms. Thus, success could be the creation of trusting relationships, the creation of a shared sense of place around the border, or the elevated visibility of a particular issue being addressed through the partnership irrespective of whether any explicit goals had been achieved.

WEBINAR SUMMARY

Panelists identified a series of challenges faced by binational partnerships. Obstacles may involve the presence or (in-)visibility of an institution in one country with a lack of representation in the other, which can lead to an imbalance or lack of trust. Organizational processes that require the cross-border exchange of resources (e.g., financial, material, human) are often subject to cumbersome regulation. Asymmetries exist in organizational capacities, especially with regard to NGOs in Mexico. Effective mechanisms to address these complexities may vary from state to state and from community to community. Several themes emerged from the partnership experiences discussed in the webinar.

"Relationships Are Everything" This general idea was repeated by multiple participants in several ways: as the centrality of interpersonal trust; as the value of day-to-day subnational diplomacy in intergovernmental partnerships or of citizen-to-citizen diplomacy in civil society partnerships; and more generally as the centrality of relationships, partnerships, friendships, and trust. Trust is needed not only among partners, but also in the mutual benefits of the partnership.

Leveraging Established Frameworks to Develop Trust The building of interpersonal trust that underlies effective partnerships itself depends on several factors, one of which is the existence of formal or established frameworks. Such frameworks provide a degree of structure and predictability that can facilitate the development of personal and informal relationships, which can be especially important when facilitating relationships across differences that are usually barriers to a partnership. Established frameworks have been especially helpful in developing relationships with the private sector, business actors with whom civil society, community, and public-sector officials do not normally have much horizontal interaction.

Cross-Boundary Literacy or "*Interculturalidad*" The development of relationships across differences draws attention to the principle that *interculturalidad*, intercultural communication, along with competence, is central to an effective partnership. This principle applies to communication across ethnic and racial differences as much as across sectoral differences. Indeed, there are particularly serious challenges to partnerships across ethnic differences in the border region insofar as Indigenous groups are especially disadvantaged stakeholders. Indigenous groups are also subject to different degrees of government and social recognition and appreciation on the two sides of the border.

Partnerships involving Indigenous peoples will also be more effective when other actors can actively listen to tribal leaders and councils. As noted in a U.N. decree,[8] the "free, prior and informed consent of indigenous peoples" is key to partnership.

Leadership and an Ethic of Place Leadership is intrinsic to effective partnerships. In the cross-border context, leadership can foster trust and the creation of an ethic of place: that is, an ethic in which the border is similarly understood by actors on both sides as a shared place. The place ethic also implies that partnerships will be more effective when leadership rotates among different parties, including people on each side of the border. It is also important that the actors and leaders involved in partnerships

---

[8] More information is available at: http://www.fao.org/3/a-i6190e.pdf.

are rooted in the location of interest rather than engaging with it from a distance as absentee economic, political, or academic actors. This ethic of place is often complicated because each country's capital city is far from the border region; the webinar panelists supported the idea that narratives about the border region are best controlled by the actors rooted there.

Fostering Inclusivity Considering and respecting different knowledge systems play a key role in fostering inclusion. The process of building trust is long and gradual; partners have a responsibility to be inclusive with their stakeholders. In addition to engaging organizations, partners need to do extensive community engagement to build trust in a broader sense. Citizen-to-citizen diplomacy is essential to address sustainability challenges. Involving the private sector remains a challenge in the region. SANDAG provides an example of a partnership that implements participatory mechanisms that build public trust, yielding more leverage to engage the private sector.

Financial Resources Webinar panelists noted the importance of money for partnership development, though finance did not feature as prominently in the webinar discussions as might have been expected. Several participants noted that resources are needed for many partnership activities. How these financial resources are distributed is also significant. Symmetrical distribution of financial resources is more likely to favor effectiveness in partnerships; however, financial asymmetries are not always an obstacle. Partnerships *have* emerged in contexts where financial resources are controlled by just a few actors, and in most cases, the actors that control greater resources are typically private corporations and organizations based in the United States. Whether the asymmetries are an obstacle depends on how they are managed and on the levels of interpersonal trust among the partners. A negative example comes from mining companies in the border region, which have been perceived as sharing relatively little information with other actors about their use of resources, especially water.[9] Asymmetries of power between mining companies and other actors are also accentuated by laws that give companies preferences. (See Appendix D for more discussion on mining partnerships in the region.)

Political Factors Diverse political factors impinge on the success of partnerships. The nature of relevant political factors varied across the cases discussed at the webinar, but among those mentioned were the relative interest and support of border governors and federal authorities, as well as the degree of political cooperation among them. A relative lack of interest and support from governors can complicate partnerships, especially those

---

[9]It should be noted that all Mexican mining companies have to provide their usage of resources in a report to the government, and most post this information on their websites; however, stakeholders sometimes do not know how to find this information.

involving subnational government authorities. When federal authorities attach a negative stigma to the border area or when they recentralize to the federal-level authorities that had previously been delegated to actors closer to the border (as in the case of the Border Health Commission), it compromises the success of partnerships.

Panelists also emphasized the need for stakeholders to better understand how official government structure and decision making impact partnerships. Even when local stakeholder partnerships establish a shared vision and common agendas, engaging government actors can be a significant challenge. Nevertheless, government engagement is invariably required because political will and sustained commitment are needed structurally to bring about positive change.

Information and Knowledge Strategies Information and knowledge strategies involve improving the availability and quality of information on existing partnerships, on how they operate, and on the factors that appear to favor their effectiveness. This information would be a source of data for participants in different partnerships, allowing them to learn from other experiences, draw lessons that they could apply to their own partnership, and even forge links or build synergies across partnerships. It is best if these information bases are publicly accessible (within the constraints of legitimate information disclosure) and at least bilingual. There is also a case to be made for making some of this information available in Indigenous languages.

Learning Strategies Closely related to information strategies is the need to develop strategies that facilitate the exchange of experiences across different partnerships and that promote the possibility of critical dialogue so that learning can occur. The webinar itself was valued by participants as just such an opportunity to hear each other and learn from each other. Strategies to promote such active learning across partnerships are therefore an important ingredient for the effectiveness of individual partnerships, at least by making their members aware of other interconnected sets of problems that likely impinge on the problems they are addressing. These learning strategies can also foster greater understanding among those actors who influence the context in which partnerships thrive or dwindle. For instance, study tours of the border area for political leaders could foster greater understanding and support from these leaders. More generally, cross-border university partnerships can help bring people together in ways that both foster learning and contribute to relationship building.

Cultivating Leadership Partnership processes are slow, and if partnerships last a long time (which is one measure of effectiveness), they will experience several rounds of leadership change. Consequently, the ongoing and active cultivation of leadership across generations is important for the continued effectiveness of partnerships. The type of leadership cultivated should be oriented toward collaborative, cross-border relationships.

All organizations have a role to play in this process, though it may be that the academic community has a particular contribution to make in how it trains emerging professionals and the types of skills and values that this training imparts.

Narrative Building Alongside civic space, the existence of narratives that create favorable environments for partnerships is also critical. The construction of narratives—frequently in the face of less than favorable dominant ideas—is a long-term process. In a context in which the border is cast as violent, unruly, and of little positive interest to the centers of authority in the two countries, the success of partnerships will be enhanced if this adverse set of ideas can be reversed and recast. Building narratives about a border with potential and with a wide range of positive human, environmental, and social assets is, therefore, a critical part of a strategy of building contexts that are favorable for partnerships. Following from the point about cultivating new leadership, these narratives would be best to be crafted in terms defined by younger people, who constitute the emerging next generation of leaders. All actors have a role to play in this narrative building, though most important is that it be done in a way that is deliberate and coordinated. This narrative building is challenging because of the transient nature of large proportions of the population along the border, at least on the Mexican side. The border needs active champions and championing.

Time The success of a partnership may depend on the ability of key actors to take a long approach. This approach allows for slow interactions, the slow building of trust between, for instance, the public and private sectors, and steady learning and adaptation. Resilient partnerships are not created from one day to the next, or from one year to the next. The strategic implication is that taking time is critical, and entering into partnerships with a long-term view and patience is likely to enhance the overall success of the partnership. Although it is important to find ways within the community of researchers and government organizations to provide support for the issues that are at hand, be they health or access to freshwater, it is a long process to build partnerships that stand the test of the time. Again, the challenge is to give partnerships time to mature when working with transient communities and in the face of pressing social and environmental challenges.

Concluding Overview Anthony Bebbington (committee member) wrapped up the webinar by noting that the key challenges for effective partnership involve recognizing the scale of the border and including those actors who have been otherwise excluded by the tendency to pull power away from the region. Central to this pursuit are narratives, as was mentioned many times during the webinar, and the need to tell the story of the border as it is and not as it is constructed to be. This needs to be done in a way that involves diverse pieces of knowledge, not just singular knowledge. And "that's certainly a warning shot across the bows of academics," he said. Key to that process, and the process of building effective partnerships, is what Bebbington referred to as "interculturality," the vital importance of recognizing a broad range of kinds of knowledge and being able to find a way to communicate across them, which may involve bridging differences between Indigenous and non-Indigenous communities or across generations or just across different organizational cultures.

KEY INSIGHTS FROM THE WEBINAR

The webinar served as a unique base of knowledge to inform the committee's subsequent analyses and deliberations. The manner in which the webinar was conducted was conducive to rich discussions in the panel session, which often spilled over into conversations in the chat room. All observed discussions converged into eight main themes, which can be summarized in no particular order as follows:

- Regardless of the sector or objective, robust communication and effective decision making are common characteristics of successful partnerships.
- Intercultural communication, patience, and cultural sensitivity are essential to establishing trust and strong relationships with local communities and are the key to understanding the effects of changes such as industrialization on Indigenous lands. It is important to inform and receive consent from Indigenous peoples about the changes that affect tribal lands.
- State and federal policies are usually at odds with border needs and priorities. Fostering co-creation at the border level is challenging in the region due to asymmetries, power dynamics, and knowledge systems (see Appendix D for more on the asymmetries and key governance challenges facing sustainable development in the region). Community-led partnerships are essential to respond to and combat the conventional, top-down approach.
- Stability is vital for binational partnerships. Government transitions on both sides of the border can affect the strength of formal partnerships. Political administration turnovers complicate processes in both countries.
- The construction of the border wall has complicated partnerships with Indigenous communities and has disrupted ecological flows. This problem is poised to grow if current tendencies in border management do not change.

- It is difficult to simultaneously empower local involvement and garner meaningful support from government agencies for binational partnerships. Because citizen participation in community processes has different histories in both countries, it is important to engage binational stakeholders in ways that do not perpetuate asymmetries.
- The environment, public health, education (including exchange visits), migration, and commerce represent the themes and sectors with the most durable partnerships. Such partnerships are particularly effective for binational communications and coordinating rapid response to acute and chronic challenges.
- Binational sustainability partnerships are in need of improved evidence and data for decision making, as well as human, financial, and institutional resources to maintain or extend their effectiveness.

4

Recommended Strategies for Effective Partnerships

Sustainable development is grounded in the protection of the natural environment, both as the functional life-support system that provides environmental goods and services, and for the preservation of its species, habitats, and complex ecosystem. The U.S.–Mexico binational region is a complex social-ecological system (SES) shaped by multiple global, regional, and local processes that are intertwined with human and ecosystem dynamics. The region faces many ongoing challenges to the sustainability of its natural resources and the livelihoods of its residents. These are exacerbated by global climate change, increasing urbanization and industrialization, and rapid population growth, as well as policy differences and diplomatic tensions that reflect national political agendas (see Appendix D). Navigating these challenges and preserving the area's cultural richness, its vibrant economy, and complex ecology will require strengthening existing—and building new—strategic partnerships that engage a broad range of stakeholders in both countries. The lack of comprehensive empirical data on sustainability partnerships in the U.S.–Mexico border region makes it difficult to obtain a complete picture of the number, type, quality, longevity, and effectiveness of the region's binational sustainability partnerships. Nevertheless, the data that exist and the accounts shared at the stakeholder webinar overwhelmingly show that effective partnership strategies and broadly acceptable metrics (Stibbe et al., 2019) that support sustainable development can enhance the well-being of the region's residents.

The United Nations Sustainable Development Goal (SDG) 17 and the 2030 Agenda Partnership Accelerator reinforce the importance of cross-sectoral and innovative multi-stakeholder partnerships in promoting "enhanced understanding of relationships across participants from different sectors."[1] As mentioned in Chapter 2, effective partnerships involve the development and application of knowledge and information, services, skills, and financial resources. Effective partnerships also require an understanding of organizational processes, including cultural and organizational values, as well as expected outcomes. A review of the current understanding of partnership effectiveness shows that developing a shared vision, iterative and participatory decision making, and knowledge co-production are fundamental to collective value creation and sustainable development implementation strategies. The webinar discussions covered in Chapter 3 underscore the importance of adopting adaptive procedures, fostering inclusivity, enhancing trust, and prioritizing sensitivity to context to enhance institutional collaboration in cross-border settings.

THE SOCIAL-ECOLOGICAL SYSTEMS APPROACH

Transborder, multi-stakeholder partnerships pursuing sustainable development are faced with highly complex and often conflicting societal and environmental goals, seeking to balance livelihoods, resource and environmental security, biodiversity conservation, and land degradation in the context of global markets and climate change. Looking at transborder partnerships for sustainability through a systems approach places equal emphasis on the social and ecological dynamics of the border region and provides insight on how to understand the important feedback each component yields to others in the system.

There are three central reasons why effective partnerships must take a systems approach. First, tackling complex problems often requires transformative system change with novel governance (Stibbe et al., 2019) and open communication. Partnerships are evolving, and adaptive processes and not fixed end-products (Stibbe et al., 2019, p. 19); the ability to adapt to such changes while continuing to provide services is a measure of SES resilience. Stakeholders in the U.S.–Mexico transboundary region are highly diverse and represent different civil and governmental interests, scientific disciplines, and knowledge bases, as well as political and cultural perspectives. They may speak different languages and approach partnerships in different and unique ways. Partners in the binational context are differentially accountable to local, regional, and national stakeholders, a fact that introduces additional complexities into partnership strategies. In addition, relationships and communication between the public and private sector, and between governmental and nongovernmental organizations, both within and across country lines, can be strained. As noted in Chapter 3, several stakeholders

---

[1] More information is available at: https://sustainabledevelopment.un.org/PartnershipAccelerator.

underscored the importance of clear and open information and communication channels to building relationships and institutionalizing input between cross-border partners. As was noted in the committee's discussion of the mining sector, some border-region industries share relatively little information with other actors about their use of resources, even when the resources are of mutual importance to several stakeholders—like water.

Second, responding to unpredictable shocks or extreme stressors, including emerging infectious diseases such as COVID-19, requires the mitigation of multiple, often unforeseen risks (Di Marco et al., 2020). Maintaining and enhancing ecosystem services to support societal well-being and equitable economic development is essential to sustainable SES development. An effective partnership between the United States and Mexico in the border region should focus on strengthening adaptability—a key SES trait—so that actors can sustain the partnership while responding to changing conditions. Chapter 3 discussed the importance of leveraging both formal and informal methods of communication to strengthen and maintain sustainable partnerships; when the pandemic eliminated in-person meetings and restricted cross-border supply chain distribution, local physicians and stakeholders relied on informal networks and virtual communication to interact with patients and to receive and send essential protocol information. Collective binational response such as this can bolster partnership viability and mitigate the adverse effects of unpredictable external disturbances.

Third, because the SDGs are interlinked and interdependent, taking an SES approach can generate synergies among sectors and actors at the highest levels of the national government and lead to integrated sustainable development plans supported by political leadership (Stafford-Smith et al., 2017). Partnerships aiming to achieve food, energy, and water security, biodiversity conservation, and climate change mitigation as key pillars of sustainable development are socially and politically acceptable means toward pursuing sustainable development that is consonant with complex SES objectives (Di Marco et al., 2020; Fu et al., 2019; Stafford Smith et al., 2017). Webinar participants agreed that although the process can be complex and challenging, engaging federal government agencies in sustainability partnerships is often a necessary step to garner sustained, widespread support and effect positive change (see Chapter 3).

In an SES, managing interdependent challenges together and closing collaborative gaps so that sustainability issues are tackled jointly may reduce emerging, often undetected, and thus unaccounted-for tradeoffs (Bergston et al., 2019). The 2018 workshop on advancing sustainability of U.S.–Mexico transboundary drylands (NASEM, 2018) highlighted the Los Alisos water treatment plant in Sonoyta, Sonora, Mexico, as an example of successful binational collaboration in which the United States and Mexico addressed a shared challenges by leveraging their respective resources. After thorough analysis of the region's water conditions, water direction, and pumping and energy needs, the North American Development Bank, U.S. Environmental Protection Agency, and Mexican government developed a plan to install solar panels at the Mexican plant. The Los Alisos wastewater treatment plant is the first in Latin America to run exclusively on solar energy.

RECOMMENDED STRATEGIES FOR FORMING AND MAINTAINING SUCCESSFUL U.S.–MEXICO BINATIONAL SUSTAINABILITY PARTNERSHIPS

Functional collaboration stresses the complementarity among partners and the recognition of the benefits of coordinated action. This is true for any transdisciplinary or transcultural (transnational) collaboration (Klein, 1996; Pohl, 2005). While every effective multi-stakeholder partnership in the U.S.–Mexico border region will not require the full representation of all groups of interest that work at the border and across the region (e.g., government, nongovernmental and nonprofit organizations, academic, business, civil society, and Indigenous communities), they can benefit from employing the following strategies.

Strategy 1: Identify Critical Issues to Be Addressed by the Partnership

It is important for stakeholders to have a clear, mutual understanding of the explicit objectives of a partnership. Developing this understanding involves identifying a target audience and location of influence for the partnership's activities while considering the impacts the partnership will have on other audiences and processes. When framing the partnership's desired outcomes, partners need to acknowledge relevant assumptions, for example, resource availability, institutional and managerial capacity, and co-dependent processes such as organizational scrutiny or political criticism, as well as the risks involved in pursuing their outcomes. A targeted focus on critical challenges and outcomes may be at odds with a more open-ended, inclusive "big tent" approach to outcome framing that itself can have inclusiveness and resource merits. Partners also need to identify tradeoffs and understand and accept that there is always uncertainty with respect to desired outcomes.

Strategy 2: Establish Trust Among Partners

Relationship building is essential to successful partnerships, often starting long before a formal partnership has been established among stakeholders and continuing well after it has ended. There is great value in practicing diplomacy within intergovernmental and civil society partnerships. However, a project's or a program's timing and a desire for efficiency and effectiveness often do not lend themselves to the pace of learning societal norms and acquiring cultural sensitivity that help foster and build partners' trust.

For U.S.–Mexico sustainability partnerships, particularly those involving representatives from Indigenous communities, *interculturalidad* (intercultural communication and competence) is a key capacity. Of particular importance are sovereign Native Nations' relationships with the United States and Mexico as other governments. Indigenous people's vision for development, goals, and objectives in building partnerships and opportunities for Native communities have historically been marginalized, even though they often have a very comprehensive understanding of the border region ecosystems.

As noted in Chapter 3, community engagement and citizen-to-citizen diplomacy aid in building trust in a broader sense. Though connecting local and governmental agencies with the private sector can prove challenging, successful cross-sectoral partnerships can work to build public trust. Developing new, beneficial relationships among stakeholders and actor groups involves establishing continuous and open dialogue, having an agreed-upon partnership structure (often involving a formal memorandum of understanding), and creating procedures for conflict resolution.

Strategy 3: Balance and Organize Power Dynamics

Achieving and maintaining successful multi-stakeholder partnerships requires the pursuit of "horizontal" interactions among partners that are fair and transparent. This may mean rotating leadership, even if the partners vary in size, organizational strength, financial standing, and other key characteristics. Addressing power asymmetries among partners requires active listening, particularly with Indigenous communities at the border, as well as awareness of the differential risks and responsibilities for each actor of engaging in partnerships. Equitable operational plans for the partnership must factor in each partner's organizational capacity and cross-partnership complementarity of resources and assets, as well as ensuring that decision making is as equitable and fair as possible. It can also be helpful to view institutional influence and social power as enabling forces in partnership execution. Partnerships *can* emerge in contexts where financial resources are controlled by just a few actors; whether the asymmetries are an obstacle depends on the how the partnership is managed and the levels of interpersonal trust that exist between partners.

Strategy 4: Establish a Stable Governance Structure

Adopting strategies for effective partnerships requires a highly flexible and adaptive collaborative structure that incorporates robust decision making and goal-oriented action. The overall approach requires strong leadership support to articulate and pursue short-, medium-, and long-term goals that set stakeholders' expectations for partnership effectiveness. Adaptive governance of multi-stakeholder partnerships entails the adoption of iterative approaches to monitoring, assessment, and interpretation of outcomes. This may require discarding or significantly modifying the original expectations, goals, projected impacts, and internal and external benefits of the partnership. Boyle, Kay, and Pond (2001) suggest that this type of transformative governance is the process of continuously targeting the collective benefits (and values) while responding to and resolving tradeoffs in the pursuit of sustainable development.

The complex sustainability context in the U.S.–Mexico cross-border region may cause governance gaps, in which stakeholders confound challenges with actors (Bergsten et al., 2019), attributing responsibility for certain outcomes to institutions or individuals who may have little control over the circumstances. Open communication, sharing of analogous experiences, and collaborative identification of responses can mitigate these situations.

Strategy 5: Agree on a Definition of Partnership Effectiveness

For partnerships to succeed, it is essential to have identified outcomes and mutual commitments by the partners to pursue these outcomes. Despite similarities across the U.S.–Mexico border, stakeholders from cities, industries, and a range of organizations in each country will invariably hold different, potentially contradictory, perspectives on partnership effectiveness. The collective process of defining specific objectives and activities may require knowledge co-generation among partners. In particular, partnerships should strive to develop a theory of change (Taplin et al., 2013) for sustainable development, identifying obstacles and avenues for progress, and rallying champions within and outside of the partnership. Institutional learning (how partnerships incorporate success and failure) is based on iterative monitoring and evaluation mechanisms. Partners may not always agree on goals, objectives, activities, indicators, underlying assumptions, and outputs (Perz, 2019); however, when this leads to "discontinuity in action or interaction" (Akkerman and Bakker, 2011), partnership leaders may initiate a learning environment, following theories of change that require individual and community engagement. While desired outcomes may evolve, mutual commitment, open communication, and a trusted process of conducting joint activities can ensure that partnership evolution brings along all partners. Explicitly adding guidelines for partners' compliance with partnership aims and activities, as well as using practical decision-making tools, can help legitimize the partnership. All partners need to be engaged in achieving partnership goals.

Strategy 6: Develop Short-, Medium-, and Long-Term Goals

Partnership strategies can be implemented over multiple timeframes. While sustainable development is a long-term goal, pursuing it requires consistent short- and medium-term efforts, which will be enhanced through partnership-based initiatives of the kind detailed in this report. Partnerships may need to focus on short-term, tactically important activities, while also articulating and pursuing longer-term cumulative success. Sometimes "effectiveness" in the eyes of the stakeholders involved may entail a partnership having to exclude certain actors. Thus, there may be a tradeoff between effectiveness and inclusion, suggesting that success is often short-lived, and may be viewed very differently by stakeholders who are external to, or who have been excluded from, the partnership. Effective partnerships require a strategy that takes account of the timing and sequence of collective and individual partners' tasks. Periodic reevaluation of the sequence of tasks can increase the probability of achieving short-term aims. Similarly, effective partnerships require flexibility in the timing and sequencing of medium-term objectives to reach broader SDGs.

Strategy 7: Establish Guidelines for Partnership Evaluation

There are three key measures for assessing partnerships: *process* (forming partnerships, setting goals, defining stakeholder roles, and conducting partnership activities); *governance* (flexibility, equity, accountability, responsiveness, transparency, and consistency among partners and external stakeholders); and *outcomes* (results in relation to goals and associated tangible factors that emerge from partnership activities). These key measures build on the central sustainability partnership characteristics identified in Chapter 2, namely *participation, collaboration* (with additional traits of inclusiveness, and leadership), and *knowledge and its co-production*. The criteria of process, governance, and outcomes are interwoven with principles for effective partnerships, chiefly, principles to guide institutional transformation, social and political power, conflict, communication, and leadership.

Process guidelines for effective partnerships start with the way clear goals are achieved, with participants and external stakeholders jointly defining the roles and responsibilities they will pursue, and where appropriate, modifying goals. Both formal and informal means of participation are important, though each must be understood, monitored, and promoted distinctly. For example, in the pandemic, informal participation temporarily gained priority. It is essential for partnership participants and leaders to be aware of, and seek to promote, equity through procedural justice to incorporate and address the needs of less dominant actors and groups.

Latent and overt forms of internal conflict can destabilize both emerging and established partnerships if not harnessed as a force for positive change, for example when legal pursuits by Indigenous communities are used to assert resource rights. The choice of leadership approaches and the establishment of checks and balances are critically important, in process terms, when leaders are themselves involved in, or may be the cause for, conflict. These final two process guidelines—navigating power and conflict—are ultimately also governance challenges.

Governance guidelines include flexibility and responsiveness, especially the ability to produce qualitatively different strategies for different approaches to partnership goals, activities, and outcomes. Co-production within partnerships (among members and leadership) and for partnerships with external stakeholders or constituents influence the quality of those partnerships, the initiatives they pursue, and the broader communities of practice they build and sustain. Additional governance guidelines for partnerships involve setting and maintaining policies and procedures, including (where necessary) legal agreements, which enhance transparency and predictability as well as improve and ensure coherence of policy and institutional aims. Outcome guidelines for a partnership, that is, the degree to which results and impacts are generated, sustained, and equitable, are perhaps the best signal to external constituents that partnerships are effective.

Given the focus of this study on SDG 17, a more nuanced appreciation of local needs and context-specific indicators of the suite of SDGs is an important consideration. For example, water-management partnerships in the binational region are crucially important to enhance water security in this arid and semi-arid region, which is confronting growing water demands for human and ecosystem needs. Additional key considerations for partnership outcomes include resources, both material and financial, as well as capacities. Partnerships' abilities to mobilize and deliver such outcomes as knowledge sharing, expertise, technologies, and financial resources are central to their pursuit of achieving sustainable development locally, in the binational region, and globally.

CONCLUDING THOUGHTS

Sustainability challenges that are addressed through binational partnerships are not unique to the U.S.–Mexico border region. They are, however, brought into sharp relief as a result of this region's social and political context, its intertwined histories, its cultural, geographical, and ecological diversity, and its shared climate vulnerability and commercial inter-dependence. Indeed, a key strategy for the effectiveness of partnership-based initiatives is the recognition and harnessing of both challenges and opportunities presented by the region's diversity and complexity.

Sustainable development in the U.S.–Mexico border region entails incremental change across a variety of inter-dependent factors. Transformative change will only be possible over the long term through an integrated approach, building on short-term steps. An integrated, medium-term approach will require partnership efforts not only to safeguard the region's unique characteristics but also to sustain partnerships themselves. Maintaining participation, collaboration and trust-building, commitment to partnership goals, and persistence with flexibility in response to changing conditions are partnership characteristics that can be strengthened through capacity building and training, as well as through the sharing of lessons learned and mutual partnership-to-partnership support. The key to sustaining partnerships is maintaining a process of continuous learning, feedback, and organizational innovation that harnesses new communication technologies and platforms, involving partners who may have historically been sidelined, and harnessing the enthusiasm and know-how of youth. In some cases, tradeoffs are required such as excluding certain actors who may view success differently than other stakeholders. In these situations, partnerships should focus on process effectiveness in reaching short- and medium-term goals instead of idealized long-term success.

Building a shared vision internally among partners and externally with stakeholders requires intentional effort and cannot be sidelined or downplayed. Leadership as a core governance competency involves identifying and pursuing common goals, navigating power dynamics and resolving conflicts (or harnessing differences for positive change), and communicating internally and externally. Finally, human, financial, and material resources must be continually mobilized, deployed, and often conserved—both for partnership effectiveness and for the broader pursuit of sustainable development.

REFERENCES

Akkerman, S.F., and Bakker, A. (2011). Boundary crossing and boundary objects. *Review of Educational Research*, 81(2), 132–169.

Bergsten, A., Jiren, T.S., Leventon, J., Dorresteijn, I., Schultner, J., and Fischer, J. (2019). Identifying governance gaps among interlinked sustainability challenges. *Environmental Science and Policy*, 91, 27–38.

Boyle, M., Kay, J., and Pond, B. (2001). Monitoring in support of policy: An adaptive ecosystem approach. In T. Munn (Ed.), *Encyclopedia of Global Environmental Change* (pp. 116–137), Volume 4. New York: John Wiley and Son.

Di Marco, M., Baker, M.L., Daszak, P., De Barro, P., Eskew, E.A., Godde, G.M., Harwood, T.D., Herrero, M., Hoskins, A.J., Johnson, E., Karesh, W.B., Machalaba, C., Navarro Garcia, J., Paini, D., Pirzl, R., Stafford Smith, M., Zambrana-Torrelio, C., and Ferrier, S. (2020). Opinion: Sustainable development must account for pandemic risk. *Proceedings of the National Academy of Sciences*, 117(8), 3888–3892.

Fu, B., Wang, S., Zhang, J., Hou, Z., and Li, J. (2019). Unraveling the complexity in achieving the 17 sustainable-development goals. *National Science Review*, 6(3), 386–388.

Klein, J.T. (1996). *Crossing Boundaries: Knowledge, Disciplinarities and Interdisciplinarities*. Charlottesville, NC: The University of Virginia Press.

NASEM (National Academies of Sciences, Engineering, and Medicine). (2018). *Advancing Sustainability of U.S.-Mexico Transboundary Drylands: Proceedings of a Workshop*. Washington, DC: The National Academies Press. doi: 10/17226/25253.

Perz, S.G. (2019). Introduction: Collaboration across boundaries for socio-ecological systems science. In S.G. Perz (Ed.), *Collaboration Across Boundaries for Socio-Ecological System Science: Experiences Around the World* (pp. 1–33). Cham, Switzerland: Springer. doi: 10.1007/978-3-030-13827-1.

Pohl, C. (2005). Transdisciplinary collaboration in environmental research. *Futures, 37*(10), 1159–1178.

Stafford-Smith, M., Griggs, D., Gaffney, O., Ullah, F., Reyers, B., Kanie, N., Stigson, B., Shrivastava, P., Leach, M., and O'Connell, D. (2017). Integration: The key to implementing the Sustainable Development Goals. *Sustainability Science, 12*, 911–919.

Stibbe, D.T., Reid, S., and Gilbert, J. (2019). *Maximising the Impact of Partnerships for the SDGs: A Practical Guide to Partnership Value Creation*. The Partnering Initiative, United Nations. Available: https://sustainabledevelopment.un.org/content/documents/2564Maximising_the_impact_of_partnerships_for_the_SDGs.pdf.

Taplin, D.H., Clark, H., Collins, E., and Colby, D.C. (2013). *Theory of Change Technical Papers: A Series of Papers to Support Development Theories of Change Based on Practice in the Field*. New York: ActKnowledge.

Appendix A

Stakeholder Information Questionnaire

LETTER

[español abajo]

Dear Colleagues,

The U.S. National Academies of Sciences, Engineering, and Medicine and the Mexican Academy of Sciences, Academy of Engineering, and National Academy of Medicine are conducting a binational study on Sustainability Partnerships in the U.S.–Mexico Drylands Region. The study focuses on identifying strategies and solutions that can strengthen ongoing binational collaboration among government, private sector, and community stakeholders.

The study committee would like to identify former and existing U.S.–Mexico partnerships and gather input to help organize a virtual public seminar to be held on July 15, 2020. We define **binational sustainability partnerships** as: "Organizations and individuals from different sectors and interest groups within the United States and Mexico, voluntarily coming together with organizations or individuals across the United States – Mexico border to address shared binational challenges and opportunities for sustainable development that isolated efforts or national initiatives would not be able to effectively accomplish." You have been identified as a representative of an organization that is currently, or has been in the past, involved in some form of U.S.–Mexico partnership as defined above.

Please complete the brief questionnaire (approx. 5 min.) at the following link: **https://tinyurl.com/ybsqywhd**.

Your contact information will be kept confidential and will only be used to notify you of the seminar, and to follow up on your answers if the committee would like to know more about your organization and partnership(s).

For more information about the study, please visit https://www.nationalacademies.org/our-work/sustainability-partnerships-in-the-us-mexico-drylands-region-a-binational-consensus-study#sectionProjectScope.

Estimadas Colegas,

Las Academias Nacionales de Ciencias, Ingeniería y Medicina de EE.UU. y la Academia Mexicana de Ciencias, Academia de Ingeniería y Academia Nacional de Medicina están realizando un estudio binacional sobre **Alianzas para la Sostenibilidad de las Tierras Áridas entre los Estados Unidos y México**. El estudio se enfocará en identificar estrategias y soluciones que puedan fortalecer la colaboración binacional en curso entre el gobierno, el sector privado y actores locales.

El comité de estudio quisiera identificar alianzas actuales y pasadas entre EE.UU. y México, y recopilar información para ayudar a organizar un seminario público virtual que se realizará el 15 de julio 2020. El comité define las alianzas binacionales de sostenibilidad como: "**Organizaciones e individuos de diferentes sectores y grupos de interés dentro del Estados Unidos y México, que se asocian voluntariamente con organizaciones o individuos a través de la frontera EE.UU. - México para abordar retos binacionales compartidos y oportunidades para el desarrollo sostenible que los esfuerzos aislados o las iniciativas nacionales no podrían lograr de manera efectiva.**" Usted ha sido identificado como un/a representante de una organización que está actualmente, o ha estado en el pasado, involucrada en alguna forma de alianza entre EE.UU. y México, como se definió anteriormente.

Por favor, complete el breve cuestionario (aprox. 5 min.) en el siguiente enlace: **https://tinyurl.com/y8gu2dj8**.

Su información de contacto se mantendrá confidencial y solamente se utilizará para notificarle a usted sobre el seminario, y para dar seguimiento a sus respuestas si el comité desea obtener más información sobre su organización y alianza(s).

Para mayor información sobre el estudio, visite a https://www.nationalacademies.org/our-work/sustainability-partnerships-in-the-us-mexico-drylands-region-a-binational-consensus-study#sectionProjectScope.

APPENDIX A

QUESTIONNAIRE

Stakeholder Information Questionnaire
Cuestionario de Información sobre Actores Locales

(Red text is instruction language for questionnaire designer; it will not appear to questionnaire users.)

(write in fields—mandatory, except for #5)
1. NAME
2. NAME OF ORGANIZATION
3. JOB TITLE
4. PHONE
5. WHATSAPP (optional)
6. EMAIL

1. NOMBRE
2. NOMBRE DE LA ORGANIZACIÓN
3. NOMBRE DEL PUESTO
4. NUMERO DE TELÉFONO
5. WHATSAPP (optional)
6. CORREO ELECTRÓNICO

7. IS YOUR ORGANIZATION NON-PROFIT OR FOR-PROFIT? (Select one)
    - NON-PROFIT
    - FOR-PROFIT

   ¿ES SU ORGANIZACIÓN SIN FINES DE LUCRO O CON FINES DE LUCRO? (Select one)
    - SIN FINES DE LUCRO
    - CON FINES DE LUCRO

8. IF NON-PROFIT, PLEASE SELECT THE TYPE: (Select one)
    - FEDERAL GOVERNMENT
    - STATE OR LOCAL GOVERNMENT
    - NON-GOVERNMENTAL
    - NOT APPLICABLE—FOR-PROFIT ORGANIZATION

SI NO TIENE FINES DE LUCRO, POR FAVOR SELECCIONE EL TIPO: (Select one)
- GOBIERNO FEDERAL
- GOBIERNO ESTATAL O LOCAL
- NO GUBERNAMENTAL
- NO APLICABLE - ORGANIZACIÓN CON FINES DE LUCRO

9. WHICH SECTOR BEST DESCRIBES YOUR ORGANIZATION? (Select all that apply)
   - MIGRATION
   - CLIMATE CHANGE / ENVIRONMENTAL CONSERVATION
   - DISASTER / EMERGENCY MANAGEMENT
   - CRITICAL RESOURCE MANAGEMENT (WATER/ENERGY/FOOD)
   - MINING / EXTRACTION
   - ARTS / CULTURE / PRESERVATION
   - EDUCATION / RESEARCH
   - ENVIRONMENTAL JUSTICE
   - URBAN PLANNING AND DEVELOPMENT
   - PUBLIC HEALTH
   - TRADE/COMMERCIAL MANUFACTURING
   - TRANSPORTATION
   - HUMANITARIAN AID
   - OTHER [Describe]

   ¿CUÁL SECTOR DESCRIBE MEJOR SU ORGANIZACIÓN? (Seleccione todas las que correspondan)
   - MIGRACIÓN
   - CAMBIO CLIMÁTICO / CONSERVACIÓN AMBIENTAL
   - GESTIÓN DE DESASTRES Y EMERGENCIAS
   - GESTIÓN DE RECURSOS CRÍTICOS (AGUA / ENERGÍA / ALIMENTOS)
   - MINERÍA / EXTRACCIÓN
   - ARTE / CULTURA / CONSERVACIÓN
   - EDUCACIÓN / INVESTIGACIÓN
   - JUSTICIA AMBIENTAL
   - PLANIFICACIÓN Y DESARROLLO URBANO
   - SALUD PÚBLICA
   - COMERCIO / FABRICACIÓN COMERCIAL
   - TRANSPORTE
   - AYUDA HUMANITARIA
   - OTROS [Describa]

10. IN WHICH MEXICAN AND/OR U.S. STATE(S) DOES YOUR ORGANIZATION OPERATE? (Select all that apply)
    - ARIZONA
    - BAJA CALIFORNIA
    - CALIFORNIA
    - CHIHUAHUA
    - COAHUILA
    - NEW MEXICO
    - NUEVO LEON
    - SONORA
    - TAMAULIPAS
    - TEXAS
    - OTHER [Please List] _____

    ¿EN QUÉ ESTADO(S) MEXICANO Y / O ESTADOUNIDENSE OPERA SU ORGANIZACIÓN? (Seleccione todos los que apliquen)
    - ARIZONA
    - BAJA CALIFORNIA
    - CALIFORNIA
    - CHIHUAHUA
    - COAHUILA
    - NUEVO MÉXICO
    - NUEVO LEÓN
    - SONORA
    - TAMAULIPAS
    - TEXAS
    - OTROS [Por Favor Liste] _____

11. BASED ON THE ABOVE DEFINITION OF PARTNERSHIPS, HAS YOUR ORGANIZATION PARTNERED BINATIONALLY WITH OTHER ORGANIZATIONS, EITHER NOW OR IN THE PAST? (Select one)
    - YES
    - NO

    BASADO EN LA DEFINICIÓN ANTERIOR DE ALIANZAS, ¿SE HA ASOCIADO SU ORGANIZACIÓN BINACIONALMENTE, AHORA O EN EL PASADO, CON OTRAS ORGANIZACIONES?
    - SÍ
    - NO

12. WHAT IS THE AVERAGE LENGTH OF THE BINATIONAL PARTNERSHIP(S)? (Select all that apply)
    - LESS THAN ONE YEAR
    - ONE TO FIVE YEARS
    - MORE THAN FIVE YEARS

    ¿CUÁL ES LA DURACIÓN MEDIA DE LA(S) ALIANZA(S) BINACIONAL(ES)? (Seleccione todos que apliquen)
    - MENOS DE UN AÑO
    - DE UNO A CINCO AÑOS
    - MÁS DE CINCO AÑOS

---

13. WITH WHICH SECTORS HAS YOUR ORGANIZATION PARTNERED BINATIONALLY? (Select all that apply)
    - MIGRATION
    - CLIMATE CHANGE / ENVIRONMENTAL CONSERVATION
    - DISASTER / EMERGENCY MANAGEMENT
    - CRITICAL RESOURCE MANAGEMENT (WATER/ENERGY/FOOD)
    - MINING / EXTRACTION
    - ARTS / CULTURE / PRESERVATION
    - EDUCATION / RESEARCH
    - ENVIRONMENTAL JUSTICE
    - URBAN PLANNING AND DEVELOPMENT
    - PUBLIC HEALTH
    - TRADE/COMMERCIAL MANUFACTURING
    - TRANSPORTATION
    - HUMANITARIAN AID
    - OTHER [Describe]

    ¿CON QUÉ SECTORES TIENE SU ORGANIZACIÓN ALIANZAS BINACIONALES? (Seleccione todos los que apliquen)
    - MIGRACIÓN
    - CAMBIO CLIMÁTICO / CONSERVACIÓN AMBIENTAL
    - GESTIÓN DE DESASTRES Y EMERGENCIAS
    - GESTIÓN DE RECURSOS CRÍTICOS (AGUA / ENERGÍA / ALIMENTOS)
    - MINERÍA / EXTRACCIÓN
    - ARTE / CULTURA / CONSERVACIÓN
    - EDUCACIÓN / INVESTIGACIÓN
    - JUSTICIA AMBIENTAL
    - PLANIFICACIÓN Y DESARROLLO URBANO

APPENDIX A

- SALUD PÚBLICA
- COMERCIO / FABRICACIÓN COMERCIAL
- TRANSPORTE
- AYUDA HUMANITARIA
- OTROS [Describa]

14. HOW WOULD YOU CHARACTERIZE THE EFFECTIVENESS OF THE BINATIONAL PARTNERSHIP ACTIVITIES IN PURSUING YOUR ORGANIZATION'S GOALS? (Select one)
    - HIGHLY EFFECTIVE
    - MODERATELY EFFECTIVE
    - MINIMALLY EFFECTIVE
    - NOT EFFECTIVE

    ¿CÓMO CARACTERIZARÍA LA EFFECTIVIDAD DE LAS ACTIVIDADES DE LA (LAS) ASOCIACIÓN(ES) BINACIONAL(ES) PARA LOGRAR LOS OBJETIVOS DE SU ORGANIZACIÓN? (Select one)
    - MUY EFECTIVA
    - MODERADAMENTE EFECTIVA
    - MINIMAMENTE EFECTIVA
    - NO EFECTIVA

Appendix B

Webinar Agenda

Binational Consensus Study on Sustainability Partnerships in the U.S.–Mexico Drylands Region

VIRTUAL PUBLIC SEMINAR ON SUSTAINABILITY
PARTNERSHIPS IN THE U.S.–MEXICO DRYLANDS REGION
Wednesday, July 15, 2020
12 pm–4 pm

Simultaneous translation available in English and Spanish

AMPLIFYING VISIBILITY

12:00 pm   **Welcome and Introduction**
José Luis Morán, *President, Academia Mexicana de Ciencias*
Toby Warden, *National Academies of Sciences, Engineering, and Medicine Board on Environmental Change and Society (BECS) Director*
Jordyn White, *Consensus Study Director*
Christopher Scott, *Consensus Study Chair*

12:20 pm   **Overview of the Seminar and the Committee's Work to Date**
Christopher Scott, *Chair*
Natalia Martínez Tagüeña, *Committee Member*
A brief recap of the goal of the consensus study, including the committee's definition of binational partnership, laying out expectations for today's event and how it will contribute to the final report.

12:40 pm   **A Look at the Landscape of Current Partnerships**
Moderator – Exequiel Rolón, *Committee Member*
Discussants –
Andy Carey, *Border Philanthropy Partnership*
Irasema Coronado, *Arizona State University*
Zachary Hernandez, *San Diego Association of Governments*

Discussants will be asked to address:
Based on the committee's understanding of binational partnerships, outlined in the past sessions, please comment on how your initiative:
- engages with and involves organizations from Mexico and the U.S.
- is connected formally or informally around key sustainability challenges
- explores existing gaps and opportunities within these partnerships.

1:40 pm   Break

1:50 pm   **Stakeholders' Approach to Successful Partnerships**
Moderator – Alma Cota de Yanez, *Committee Member*
Discussants –
Benjamin Wilder, *Next Generation Sonoran Desert Researchers (N-Gen)*
Trevor Hare, *Watershed Management Group*
Yoselin Cárdenas, *Consejo Empresarial Nogales, A.C.*
James Callegary, *U.S. Geological Survey*

Discussants will be asked to describe and characterize:
- ways in which partners communicate with each other and their audiences, including in the COVID-19 pandemic
- systematic approaches you follow to address new opportunities for consolidation or expansion of your partnerships
- the use data and scientific evidence to support decision-making.

APPENDIX B

2:50 pm   **Breaking the Silos: Challenges to Successful Partnerships**
Moderator – Hallie Eakin, *Committee Member*
Discussants –
Gabriela Múñoz Meléndez, *El Colegio de la Frontera Norte*
Blake Gentry, *Policy Advisor - Lideres Tradicionales de O'odham en Mexico*
Octaviana Valenzuela Trujillo, *Northern Arizona University*
Gabriel Armenta, *Índex Nogales. Asociación de Maquiladoras de Sonora, A.C*

Discussants will be asked to describe and characterize:
- *how you identify obstacles in cross-country and cross-sector collaboration (visibility, organizational sizes, and structures, domestic and international regulation, trust, etc.)*
- *systematic approaches you follow to address challenges for your partnership activities, including in the COVID-19 pandemic.*

3:50 pm   **Wrap-Up; Committee's Next Steps**
Anthony Bebbington, *Committee Member*

4:00 pm   Adjourn

Appendix C

Committee Member and Staff Biographies

Christopher A. Scott (*Chair*) is the director of the Udall Center for Studies in Public Policy and research professor of water resources policy, with joint appointments as professor in the School of Geography and Development at the University of Arizona as well as director of the Consortium for Arizona–Mexico Arid Environments and joint professor of hydrology and atmospheric sciences in the College of Science. In 2021, he moved to Pennsylvania State University, where he was named Maurice Goddard Chair of Forestry and Environmental Conservation and professor in the Department of Ecosystem Science and Management. His work focuses on the policy dimensions of global climate change and urban growth, with particular emphasis on water and energy security, climate adaptation, urban wastewater and water reuse, agricultural-urban water transfers, and transboundary water resources. He is founding codirector of the AQUASEC Center of Excellence for Water Security. He has bilingual proficiency in Spanish and Hindi, and professional working proficiency in Portuguese, Nepali, and German. Prior to joining the University of Arizona, he was a senior international project manager with the National Oceanic and Atmospheric Administration, where he led the National Weather Service collaboration with Mexico and India. Scott holds a Ph.D. and an M.S. in hydrology from Cornell University, and B.S. and B.A. degrees from Swarthmore College.

Anthony Bebbington is international director for Natural Resources and Climate Change at the Ford Foundation, Milton P. and Alice C. Higgins Professor of Environment and Society in the Graduate School of Geography at Clark University, USA (on leave), and professorial fellow at the

University of Manchester, UK. His research and teaching focuses on environmental governance, socioenvironmental conflicts, resource extraction, and community rights, primarily in Latin America. He is a member of the National Academy of Sciences and the American Academy of Arts and Sciences, a director of Oxfam America, and a distinguished professor at the Latin American Faculty of Social Sciences, Ecuador. He has been a Guggenheim fellow, an Australian Research Council Laureate fellow, fellow at the Center for Advanced Studies in the Behavioral Sciences, at Stanford University, and a social scientist at the World Bank. Additionally, he has served on numerous National Academies of Sciences, Engineering, and Medicine panels. Recent books include *Governing Extractive Industries: Politics, Histories, Ideas*, *Subterranean Struggles: New Dynamics of Mining, Oil and Gas in Latin America*, and the collection *Impacts of Extractive Industry and Infrastructure on Forests, Climate and Land Use Alliance*. Bebbington earned his Ph.D. in geography from Clark University.

Alfonso Andrés Cortez-Lara has been a tenured professor and researcher at El Colegio de la Frontera Norte since 1993. He is a member of the System of National Researchers Level II (SNI II-Conacyt). His most recent book covers the issue of transboundary water conflicts in the Lower Colorado River Basin. He is currently a cosponsor researcher for two studies: Democratización de las instancias de toma de decisiones sobre aguas y cuencas en México, which concerns water and basins in Mexico, and Los trasvases como dispositivos de desigualdad e inseguridad hídrica; Prácticas colectivas para la justicia hídrica, which concerns water inequality and water justice. Cortez-Lara has a Ph.D. in resource development (water resources management) from Michigan State University.

Alma Cota De Yañez is the executive director for Fundación del Empresariado Sonorense, A.C., Nogales (FESAC). Since assuming this position in 2003, she has helped FESAC become the local leader in mobilizing resources and philanthropy efforts for individual and corporate donors, government agencies, and nongovernmental organizationa (NGOs) providing support to improverished communities in the border towns of Nogales, Mexico, and Nogales, Arizona. She began working with NGOs as a part-time translator for Save the Children during her university studies. She graduated from the Global Women's Leadership Network international training program in 2005, and in 2007 completed an international senior fellowship with the Center on Philanthropy and Civil Society. Throughout her career, she has worked to provide training programs, nutritional services, self-employment guidance, and support for people with disabilities, often working with migrant workers and their families. Cota de Yañez has a bachelor's degree in business administration from the Technological Institute of Monterrey.

Hallie C. Eakin is a professor at Arizona State University in the School of Sustainability and an affiliated professor in the School of Urban Planning and Geographical Sciences as well as the School for the Future of Innovation in Society. Eakin's research interests include household vulnerability and the sustainability of adaptations to global change, social-ecological resilience and transformation, urban resilience planning and governance, social justice concerns associated with global change, the governance of telecoupled systems, sustainable food systems, agricultural change, and food sovereignty. Eakin's most recent work has explored the implications for social vulnerability of water infrastructure decision making in Mexico City, a National Science Foundation project implemented in collaboration with the Laboratory for Sustainability Sciences of the Universidad Nacional Autónoma de México in Mexico City. She received her Ph.D. in geography from the University of Arizona and completed postdoctoral fellowships at the U.S.–Mexican Studies Center, University of California-San Diego, and the Centro de Ciencias de la Atmósfera, Universidad Nacional Autónoma de México.

Constantino de Jesús Macías Garcia is a professor and former director (2016–2020) of the Instituto de Ecología, Universidad Nacional Autónoma de México (UNAM) where he oversees the work of the National Laboratory for Sustainability Sciences and has been involved in proposals to strengthen collaboration between his university and the universities of Arizona and Agadir (Morocco) to work on arid-land sustainability. His main research area has been evolution through sexual selection and its impact on the generation of new species, focusing on Mexican native fish, but increasingly he has been working to understand how animals adapt to habitats modified by humans where the main study system has been urban birds, working on how they adapt their song attributes/behavior to urban noise, and how they use anthropogenic materials to build their nests and with what consequences. He is the associate editor of *Behavioral Ecology and Sociobiology* and is a referee for many scientific journals. Macías Garcia is a member of the Mexican Academy of Sciences. He earned a B.Sc. and M.Sc. in biology at the UNAM School of Sciences and a Ph.D. from the University of East Anglia, Norwich, in animal behavior.

Natalia Martínez Tagüeña is an environmental anthropologist and archaeologist doing research at the Consorcio de Investigación, Innovación y Desarrollo para las Zonas Áridas, located at the Instituto Potosino de Investigación Científica y Tecnológica in San Luis Potosí, México. The goal of the partnership in the institute is to conduct transdisciplinary and participatory research for the sustainable use of natural resources in arid lands. Her research interests have a regional focus on arid lands, particularly at the Sonoran Desert, while studying diverse topics including the employment of past information to better understand subsistence and climate change today, transitions from mobile to sedentary lifeways, coastal adaptations, ethno-ecology, traditional knowledge, cultural landscapes, and socio-ecological systems. She conducts community-based and participatory research enriched by collaboration, where different epistemologies are integrated to achieve co-produced knowledge. Martínez Tagüeña received her Ph.D. in anthropology from the University of Arizona in Tucson.

Roger S. Pulwarty is a senior scientist at the National Oceanic and Atmospheric Administration (NOAA) Earth System Research Laboratory Physical Sciences Division in Boulder, Colorado, as well as adjunct professor to the University of Colorado and the University of the West Indies, Barbados. He has extensive experience working with Native American communities and protected areas in the Southwest United States/Northwest Mexico region, and on transboundary water resources and research networks to support adaptive management on the Colorado River between the United States and Mexico. Pulwarty has developed and led multidisciplinary programs for the U.S. National Integrated Drought Information System; NOAA Regional Integrated Sciences and Assessments; the World Meteorological Organization Climate Services Information System; and the Global Environment Facility Mainstreaming Adaptation in the Caribbean. He is a fellow of the American Meteorological Society and of the American Indian Science and Engineering Society. He is the coeditor of *Drought and Water Crises*, *Drought in the Anthropocene*, and the U.N. Office for Disaster Risk Reduction's special global assessment report on drought. Pulwarty earned his Ph.D. in geography from Colorado University, Boulder.

Exequiel Rolón is sustainability manager of Fresnillo PLC, the world leader in silver mining, where he is responsible for the social performance of the company. He works closely with operations and development projects to engage and build trust with neighboring communities. In addition, he manages sustainability reporting and leads the initiatives to foster diversity and embed ethics in the organizational culture. Prior to Fresnillo PLC, he was a consultant on projects in Canada, Peru, the Dominican Republic, and Madagascar. He currently serves as a board member of the Center for Leadership Ethics of the University of Arizona and the World Environment Center. He also participates in the sustainability initiative of the Silver Institute and regularly speaks at conferences and events on sustainability and community relations. Rolón received his B.S. in civil engineering from Universidad Panamericana, an M.Sc. in geomatics from Université Laval, and an M.B.A. from HEC Montreal.

Kelly T. Sanders is an associate professor in the University of Southern California's Sonny Astani Department of Civil and Environmental Engineering. Her research aims to ease tensions between human and natural systems through technical, regulatory, and market interventions, with particular emphasis on reducing the environmental impacts of providing energy and water services. She has authored more than two dozen publications and has given dozens of invited talks on topics at the intersection of engineering, science, and policy. Sanders has been recognized in *Forbes*' "30 under 30 in Energy" and *MIT Technology Review*'s "35 Innovators Under 35" for her contributions to the energy field. Sanders received her B.S. in bioengineering from Pennsylvania State University, as well as M.S.E. and Ph.D. degrees in mechanical engineering and environmental engineering from the University of Texas at Austin, respectively.

Elisabeth Huber-Sannwald is a research professor in the Division of Environmental Sciences at the Instituto Potosino de Investigación Científica y Tecnológica in San Luis Potosi, Mexico. Her expertise is in dryland ecosystem ecology with a focus on diversity and the functioning of plants, biocrusts, soil microorganisms and their role in ecohydrological and biogeochemical processes under the influence of global and social changes. Over the past 17 years, her studies have addressed the mechanisms underpinning the integrity of socio-ecological systems and the ways dryland resilience is linked to the nexus of ecosystem services, human well-being, and sustainable development. Her research is inter- and transdisciplinary spreading across a complex systems approach, participatory research, field experimentation, and long-term socioecological monitoring. She is the founder and coordinator of the International Network for Drylands Sustainability (RISZA), which is funded by Consejo Nacional de Ciencia y Tecnología. Currently, she is jointly coordinating the network of Socio-ecological Participatory Observatories in Mexican drylands with RISZA's Technical Academic Committee to foster community learning with local multistakeholder partnerships. After her graduate work, she served as a research assistant at the Technical University of Munich and held a postdoctoral position at the Institute of Ecology, University of Buenos Aires. Huber-Sannwald holds a Ph.D. in ecology from Utah State University.

Toby Warden is the director for both the Board on Environmental Change and Society and the Board on Human-Systems Integration at the National Academies of Sciences, Engineering, and Medicine. She oversees a range of social science-related activities concerning the human dimensions of environmental change and optimizing organizational performance. She joined the National Academies in 2009 as a study director on climate change and weather-related activities with the Board on Atmospheric Sciences and

Climate. In 2011, she joined the Division of Behavioral and Social Sciences and Education as a study director, and later associate board director, working on projects related to worker safety, safety culture, and systems science. From 2014 to 2015, she served as director of Scientific Administration for the Department of Neurological Sciences and as assistant professor at the University of Nebraska Medical Center, spearheading strategic planning efforts to foster research collaboration across the university system. Warden has a Ph.D. in social ecology with an emphasis on environmental analysis and design from the University of California, Irvine, and a certificate in business fundamentals from HBX/Harvard Business School.

Jordyn White (*Study Director*) is a program officer in the National Academies of Sciences, Engineering, and Medicine's Division on Behavioral and Social Sciences and Education. She recently directed a groundbreaking study that assessed the current state of data on the well-being of LGBTQI+ populations. Her previous projects include a study on the National Assessment of Educational Progress and workshops on estimating human trafficking in the United States and on principles and practices of federal program evaluation. Previously, at the U.S. Census Bureau, she worked on methodology, implementation, and nonresponse follow-up design for the American Community Survey and the 2020 Census. She is a member of the Advisory Committee to the Office of LGBTQ Affairs in the Office of the Mayor of the District of Columbia. White has a B.S. in psychology from the University of Pittsburgh and an M.S. in criminal justice from St. Joseph's University.

Adam K. Jones is a research associate in the National Academies of Sciences, Engineering, and Medicine's Division on Behavioral and Social Sciences and Education. He served as a senior program assistant for both the Board on Environmental Change and Society and the Board on Human-Systems Integration supporting the consensus study on Sustainability Partnerships in the U.S.–Mexico Drylands Region and the Committee on Cybersecurity Workforce of the Federal Aviation Administration. Before joining the National Academies, he served on the board of the Graduate English Organization at the University of Maryland (UMD), College Park, as the technology chair from 2018 to 2019. Jones holds an M.A. in English language and literature with a certificate in critical theory from UMD, College Park, where his scholarship focused on 20th century and contemporary literature depicting climate change and envrionmental ruin; and he received his B.A. in English literature from the University of Utah.

Tina M. Latimer is the program coordinator for the Board on Environmental Change and Society and the Board on Human-Systems Integration. She joined the National Academies of Sciences, Engineering, and Medicine in 2014 after 19 years of experience working in law firms as an office manager and executive legal secretary. She also worked as a staff assistant to the U.S. Congressional Subcommittee on Commerce, Consumer Protection and Competitiveness. Through these experiences, she developed excellent skills in managing the overall administrative and logistical procedures in a busy environment. She is responsible for coordinating the reporting requirements, administrative functions, and logistical support for both boards, the director, and the project committees. Latimer holds a B.S. in criminology and criminal justice (with a minor in women's studies) from the University of Maryland.

Heather Kreidler is the owner of Fact or Fiction, LLC, a consulting business dedicated to fact-checking services and research support. From 2008 to 2019 she worked at the National Academies of Sciences, Engineering, and Medicine with the Board on Environmental Change and Society, Board on Human-Systems Integration, Board on Children, Youth, and Families, and Food and Nutrition Board. Ms. Kreidler received a B.S. in business management from Kutztown University and an M.S. in environmental science and policy from George Mason University.

Appendix D

Characteristics of the Binational Region

Sustainability partnerships in the U.S.–Mexico binational region are products of, and aim at responding to, their regional context. While the report addresses partnerships, this appendix seeks to set the stage by describing the region's biophysical environment—its water, climate, land, and ecology—and by identifying the most salient socioeconomic forces, such as population, migration, urban growth, and various economic sectors. The concept of sustainability is a critical undercurrent when describing the binational region's diversity and evolving priorities; it provides a rich contextual background for researchers on the breadth and depth of its emerging and persistent challenges. Partnerships for sustainability, therefore, are called to consider the amplitude, complexity, and critical importance of the U.S.–Mexico region. This appendix is also intended to serve as a primer for current and future stakeholders to have a resource for understanding the complexity, and the critical importance of the context, of this region.

OVERVIEW OF THE REGION

The U.S.–Mexico border is one of the world's longest borders, spanning an estimated 1,933 miles east to west (Beaver, 2007) and 62.5 miles north to south of the international boundary.[1] Despite containing several economic asymmetries, this region, home to approximately 15 million people,

---

[1] As defined by the La Paz Agreement U.S. Department of Health and Human Services [HHS], 2017).

is also one that contains shared history, culture, environmental, and security relationships (HHS, 2017; Giner et al., 2019).

According to 2019 U.S. Census Bureau data, 7.5 million people live in the four border states (California, New Mexico, Arizona, and Texas), and 7.1 million live in the 37 Mexican municipios spanning the border. By 2025, it is expected that the border region's population will double, mostly in urban regions (HHS, 2017; Wilder et al., 2013).

The vast majority (approximately 90%) of border region populations inhabit metro areas (HHS, 2017). The majority of the border's urban centers are sister or "mirror" cities, having a counterpart directly across the border. These mirror cities (e.g., San Diego, California, and Tijuana, Baja California, San Luis, Arizona, and San Luis Rio Colorado, Sonora) are the focal point of most of the border's economic and social activities.[2] Although the cities on each side of the border have obvious similarities to those on the opposite side due to common climate and natural resources, they differ greatly in their infrastructure, resource management, legislation, culture, and language. The sprawling urbanization has been incentivized by cheap available land adjacent to existing cities, as well as the low costs of transportation and development (Giner et al., 2019).

Many of the communities on either side of the border are among the poorest and most under-resourced of any region within their respective countries, and rampant and unplanned urbanization has put great pressure on the infrastructure and natural resources supporting these communities (Giner et al., 2019). The region's mutual social, economic, and environmental priorities underscore the need for binational cooperation.

Much of the binational region is characterized by high aridity and high temperatures (Wilder et al., 2013). About half of its precipitation tends to fall in summer months (except in California), in brief, but high-intensity heavy-rain events. However, there is significant inter-annual and multi-decadal variability in precipitation patterns, which adds complexity to managing the region's scarce water resources (Giner et al., 2019; Wilder et al., 2013). Most of the arid and semi-arid regions receive well below 500 millimeters (20 inches) of rain annually, with some hyper-arid areas, such as the desert region adjacent to Yuma, Arizona, receiving less than 75 millimeters (3 inches) annually. Water scarcity across much of the region has been exacerbated by large increases in population, agricultural intensification, growth in the industrial sector, and climate change (Díaz-Caravantes and Wilder, 2014). Over time, the numbers and intensity of extreme events, such as flooding, have increased, due to climate change. These events have

---

[2]For an example of growth since 1990 in mirror cities located in Texas, Chihuahua, Coahuila, and Tamaulipas, see: https://www.tceq.texas.gov/border/population.html.

been costly in terms of damages and have been exacerbated by insufficient and poorly planned infrastructure (Giner et al., 2019).

In 1994, several binational initiatives went into effect, in concert with the North American Free Trade Agreement (NAFTA),[3] to provide financing and resources for the planning, development, and implementation of environmental infrastructure intended to protect and improve the shared environment and well-being of the residents in the border region. These efforts, financed by the North American Development Bank (NADB), have been developed in their technical, environmental, and social aspects by the Border Environment Cooperation Commission (BECC) (Congressional Research Service [CRS], 2020). The BECC's authority spans approximately 60 miles north and 185 miles south of the border through a region encompassing 13.9 million and 26.1 million residents in the United States and Mexico, respectively (Giner et al., 2019). Binational cooperation through these initiatives, in partnership with Mexico's National Water Commission, Comisión Nacional del Agua (CONAGUA), Mexico's Ministry of the Environment, Secretaría del Medio Ambiente y Recursos Naturales (SEMARNAT), and the U.S. Environmental Protection Agency (EPA), has yielded projects that have improved basic infrastructure, including improved access to drinking water, treatment of wastewater flows, and improved management of air quality, and solid waste (Giner et al., 2019).

SOCIAL SYSTEMS

The binational region is characterized by the ebb and flow of human and ecological processes across what is now the U.S.–Mexico border. Many interactions predate the border itself, though there has been a historical hardening of the border line, resulting from national policies on immigration, trade, health, and other binational exchanges.

Indigenous Communities

Along the U.S.–Mexico border there are approximately 60 tribal nations and Indigenous communities with more than 40,000 inhabitants, occupying territories in California, Arizona, and Texas in the United States, and in the Mexican areas of Baja California, Sonora, and Coahuila, respectively (Southwest Center for Environmental Research and Policy [SCERP], 2004). Though the border has split the communities in two, many of the tribal nations still maintain close cross-border relations. The Kikapú ("Kickapoo" in the United States), Kumiai (Kumeyaay), Papago (Tohono O'odham),

---

[3] More information is available at: https://www.cbp.gov/trade/nafta.

Cucapá (Cocopah), and other mobile populations, such as the Yaquis, Pima, Paipai, and Kiliwa, still maintain cultural, economic, and political ties with their cross-border counterparts (SCERP, 2004). In addition to maintaining cultural and family ties, groups such as the Kikapú engage in commerce on both sides of the border such as livestock and agricultural ranches, and some tribes also operate casinos (SCERP, 2004).

Over time, the Indigenous presence in the border area has undergone substantial changes. In the northern states of Mexico, the Indigenous population increased substantially between 1970 and 2000, a situation that modified the landscape of the region by incorporating new languages and transforming socio-cultural dynamics both within each Indigenous migrant group and between the groups and existing local populations (Rodríguez, 2016). Waves of Indigenous groups from southern Mexico migrated to form communities along the border, concentrating heavily in areas such as Chihuahua and Baja, California, and bringing with them their southern Mexican dialects; two of the most widely spoken Indigenous languages in the border region are Mixteco and Nahuatl.

Migration

The region is also populated by a large number of people seeking passage to the United States from Mexico and elsewhere in Central and South America. Because it often takes a migrant time to attempt to raise the necessary resources to pay for an illegal border crossing (without sufficient documentation), these northward migrants form a substantial "floating population" (Peña Muñoz, 2018). Floating populations also include deported individuals who are having difficulty returning to the United States or their places of origin in Mexico (Peña Muñoz, 2018), as well as migrants residing in temporary shelters or provisional encampments under the Migrant Protection Protocols.[4]

According to data from the National Institute of Statistics and Geography, Instituto Nacional de Estadística, Geografía e Informática (INEGI), the floating population in 2015 accounted for a significant fraction of the population in several border urban centers: 33 percent in Nuevo Laredo, 23 percent in Tijuana, and 27 percent in Nogales (Peña Muñoz, 2018, p. 88). Large-scale temporary migration directly impacts the local economies of those cities, enlarging the labor force by temporarily integrating into the maquiladora workforce, as well as becoming consumers of services such as lodging, food, and transport. Since early 2019, a significant number of migrants have resided in temporary shelters or provisional encampments

---

[4] More information is available at: https://fas.org/sgp/crs/row/R41576.pdf.

in Mexico's northern border states of Baja California, Chihuahua, and Tamaulipas under the Migrant Protection Protocols.[5]

Migration rates of Mexican-born individuals into the United States grew exponentially between 1980 and 2010 (Israel and Batalova, 2020). In 2010, 30 percent of all legal migrants into the United States were of Mexican origin. Between 2010 and 2019, the number of Mexican emigrants in the United States decreased by almost 780,000 and now comprises just under one-quarter of the foreign-born population. This decrease was due in part to changes in U.S. immigration policy and increased enforcement and in part to a strengthening Mexican economy (Israel and Batalova, 2020). Between 2012 and 2016, 53 percent of migrants who were undocumented in the United States were of Mexican origin; by comparison, El Salvador, Honduras, China, and Guatemala together accounted for 14 percent (Batalova et al., 2020). Most of these migrants crossed by land.

Border movements in both directions can generate temporary or permanent economic opportunities for residents in the area. These movements also translate into an increased exchange of materials and products between the two countries. In this regard, it is worth noting the importance of the urban corridors, which operate as axes along which countless cross-border interactions occur, generating economic interdependence (National Academies of Sciences, Engineering, and Medicine [NASEM], 2018). Tourism flowing from both countries to the border region has also been important in the development of the region and the service industry on both sides: according to the Border Governors Conference, an estimated 64.5 million tourists visited Mexico by car or by foot in 2019 (Secretaría de Turismo, 2019). Total vehicular movement across the U.S.–Mexico border neared 530 million in 2019.

Safety and Security

A lack of adequate human security characterizes the region. These inadequacies plague economic, food, health, environmental, community, political, and personal security (NASEM, 2018), all of which are needed as basic protections in the face of growing violence, child labor, poverty, and other social stresses. The inconsistent availability of these forms of securities and protection adds to the complexity of the border region's social, political, economic, and ecological landscape. In this context, illegal economic activities at the border, such as drug production, transfer, and sale, become a significant and pervasive issue, fostering informal economies and money laundering through the acquisition of land, hotels, and a

---

[5] More information is available at: https://fas.org/sgp/crs/row/R41576.pdf.

range of real estate. For example, in 2014 in the Benjamin Hill municipality, located south of the border in Sonora, Mexico, 12 of every 1,000 inhabitants had been convicted of drug trafficking offenses (Piña Osuna and Poom Medina, 2019). Six of Sonora's municipalities were included on Mexico's top ten list of areas nationwide with the highest rates of residents convicted of such crimes (Piña Osuna and Poom Medina, 2019). In 2018 and 2019, Ciudad Juárez and Tijuana recorded Mexico's highest incidence of violent crimes, with 1,281 murders and 2,000 homicides, respectively (Beittel, 2020). Most of the recorded homicides in Tijuana and Ciudad Juárez were connected to drug gangs, which could point to drug trafficking as playing a significant role in the social dynamic of the border region (Beittel, 2020).

Illegal trafficking of drugs from Latin America through Mexico to supply a large demand in the United States, as well as weapons and smuggled goods moving from the United States into Mexico, are persistent problems in the region. Illicit activity in the region has generated violence, corruption, and political tensions, highlighting deeper problems in both urban and rural areas, such as marginalization, poverty, and labor and social inequalities. Campbell (2007) notes that drug trafficking is socially perceived as a way to overcome poverty. For women, it presents itself as an opportunity to leave their marginalized social position (Santamaría, 2012).

Public Health Challenges

One of the main difficulties in characterizing public health conditions in the border region stems from a lack of basic data, which makes it difficult to develop comparative diagnoses and to coordinate actions based on shared health indicators (Carrillo et al., 2017). However, some binational efforts have focused on public health, such as the 2000 Mexico–U.S. Border Health Commission Strategic Plan, while binational groups, such as the Border Health Consortium of California, have focused on public health in the border region. The latter consortium meets frequently to explore opportunities for collaboration between California and Baja California. This region in particular shows some peculiarities, such as a high incidence of tuberculosis infections in both countries (Pelozzi et al., 2014).

The border region also has high rates of people without access to health insurance: while the U.S. national average is 8 percent uninsured, in Texas the rate is 25 percent, and the rate in Mexico at the border region was 34.6 percent in 2010 (Pelozzi et al., 2014). However, due to lower health care costs in Mexico's border municipalities, there is a large flow of U.S. patients crossing into Mexico to be treated. By 2018, an estimated 2.4 million foreign patients were being treated in Baja California alone.

BINATIONAL ECOLOGY

The U.S.–Mexico border traverses a region of immense ecological diversity and abundant natural resources. McCallum, Rowcliffe, and Cuthill (2014) identified the area as one of America's most biologically wealthy regions. The Commission for Environmental Cooperation, created to execute the North American Agreement on Environmental Cooperation, established by Mexico, the United States, and Canada, identified seven Level III ecoregions along the U.S–Mexico border (Wiken et al., 2011). From west to east, these ecoregions are: (1) the California coastal sage, chaparral (thicket), and oak woodlands; (2) the Sonoran desert; (3) the Madrean archipelago; (4) the Chihuahua desert; (5) the Edwards Plateau; (6) the southern Texas plains/interior plains and hills (with xerophytic shrub and oak forest); and (7) the western Gulf coastal plain. These ecoregions present various types of vegetation, from desert to grasslands to freshwater wetlands and marshes (Peters et al., 2018; Wiken et al., 2011).

In addition to being a habitat for a large number of species, south Texas' ecosystems depend on the monarch butterfly and some species of neotropical birds, which pass through on their migratory journeys (Peters and Clark, 2018). Efforts to conserve migratory avian nesting habitats here, including the binational Migratory and Shore Bird Habitat Initiative, have been planned since 2013 (Good Neighbor Environmental Board [GNEB], 2014). Other binational conservation experiences in the border region include the Big Bend Binational Initiative and the National Park Service Sister Parks Initiative (GNEB, 2014). South of the border lies the extraordinarily rich environment of Cuatro Ciénegas, a hotspot of biodiversity that represents some of the conditions that prevailed in ancient ecosystems, and has thus attracted the interest of scientists at the National Aeronautics and Space Administration and other agencies and organizations interested in astrobiology (see Pérez Ortega, 2020). In general, conservation of the region's ecology is the focus of numerous binational partnerships.

Threats to Biodiversity

In 2018, a group of more than 2,500 scientists noted that the border region contains a cumulative area of approximately 17,000 square miles (4.5 million hectares) protected through various biodiversity conservation programs, with approximately 10,000 square miles (2.6 million hectares) of sustainable-use programs (Peters et al., 2018). Four areas of protected land span the border: the Sonoran desert, Sky Islands, Big Bend, and the lower Rio Grande (Peters et al., 2018). These protected areas comprise 18 percent of the border region (Peters et al., 2018).

Various analyses of the region show that the path of the border line crosses a geographical range of 1,506 native species of terrestrial and aquatic plants and animals, including 62 that are reported as critically endangered, endangered, or vulnerable by the International Union for Conservation of Nature (Peters et al., 2018). Along the border, some organizations, such as Defenders of Wildlife, have identified five hotspots for the conservation of borderlands, which represent areas of high biological diversity that are now at risk due to the construction of the border wall (Córdova and de la Parra, 2007; Peters et al., 2018). In this regard, the Center for Biological Diversity identified 93 endangered, threatened, and candidate species that could be impacted during border wall construction, of which 88 have populations on both sides of the border and could experience a limited flow in their gene pool (Greenwald et al., 2017). This is the case for the peninsular bighorn sheep *(Ovis canadensis nelsoni)*, an endangered species, which would see its regular roaming between Mexico and California significantly limited by a wall, endangering its access to birth sites and water sources. Other species that could face the same danger are the Mexican grey wolf (*Canis lupus baileyi*), the Sonoran pronghorn (*Antilocapra americana sonoriensis*), the jaguar (*Panthera onca*), and the ocelot (*Leopardus pardalis*) (Peters et al., 2018).

CROSS-BORDER WATER FLOW

Water is one of the most consequential resources shaping social and ecological dynamics in the region. The Colorado River and the Rio Grande (known as Rio Bravo in Mexico) are the two central river systems shared by Mexico and the United States, although the Tijuana River, New River, and multiple shared aquifers also cross the border (GNEB, 2014) and provide water to residents of both countries. Precipitation rates fluctuate along the border, and this variability influences the volume of usable water in the region. For example, in Imperial Valley, California, to the west, the yearly precipitation on average is 3 inches. In the east, the precipitation varies from place to place; in 2014, average rainfall was 19 inches in Nogales, 8 inches in El Paso, and 28 inches in Brownsville, which is located at the far east end of the border, where the mouth of the Rio Grande/Bravo meets the Gulf of Mexico (GNEB, 2014).

Rivers

In the case of the Rio Grande/Bravo, its basin encompasses an area of more than 180,000 square miles and frequently runs along the international border. This river supplies water for municipal use, and its waters irrigate a combined U.S.–Mexico extension of 2 million acres (U.S. Department of the Interior, Bureau of Reclamation, 2016a). After a century of use, a change in the flow pattern is evident, and in recent decades, the amount of water has decreased (GNEB, 2014). An example of this is the Conchos River, a tributary of the Rio Grande/Bravo that joins it near the city of Ojinaga, Chihuahua. Historically, the Concho constitutes 70 percent of the flow of the Rio Grande/Bravo (Carter et al., 2017), allowing the latter to regain its water level, which in many sections before the junction drops to virtually zero. Since the 1990s, the Conchos River's contribution to the Rio Grande/Bravo has decreased, and currently, it represents only 40 percent of the total flow (Carter et al., 2017). Consequently, only a small proportion of the river's natural discharge reaches the Gulf of Mexico.

The Colorado River basin encompasses 246,000 square miles (629,000 square kilometers) across seven U.S. states and Mexico (U.S. Department of the Interior Bureau of Reclamation, 2016b). According to data from the U.S. Department of the Interior, Bureau of Reclamation (2013), "The Colorado River and its tributaries provide water to nearly 40 million people for municipal use, supply water to irrigate nearly 5.5 million acres of land, and is the lifeblood for at least 22 federally recognized tribes, 7 National Wildlife Refuges, 4 National Recreation Areas, and 11 National Parks."

The Tijuana River watershed drains 1,750 square miles (4,532 square kilometers). This basin is one of the fastest-growing regions along the border, with roughly 4.5 million people—3 million of whom live in the San Diego County area and 1.5 million in the city of Tijuana (GNEB, 2014).

Aquifers

No consensus exists on the number of cross-border aquifers between Mexico and the United States. The International Shared Aquifer Resources Management, an initiative of UNESCO, recognizes 11 such aquifers; CONAGUA in Mexico lists 36; and the 16th report of the GNEB (2014) speaks of 20 (see also Sanchez et al., 2016). However, several studies in the United States recognize the existence of at least 38 aquifers, 12 of them along the Mexico–California border, 9 along the border with Arizona, 8 along the border with New Mexico, and 9 along the border with Texas (Sanchez et al., 2016). The differences among these estimates stem from the absence of a common definition between the two countries as to the characteristics of a cross-border aquifer, as well as the lack of agreement for delimiting cross-border aquifers. Sanchez, Lopez, and Eckstein (2016, p. 8) suggest the presence of up to 36 cross-border aquifers, "albeit with different levels of confidence of their transboundary nature."

These aquifers constitute one of the largest sources of water supply in the border region and are often not addressed in binational water-sharing agreements. The most obvious and well-documented case is the binational urban complex in El Paso, Texas, and Ciudad Juárez, Chihuahua, that pumps water from the Hueco Bolsón aquifer for the 1.5 million residents of Ciudad Juárez and to 40 percent of El Paso's 730,000 residents (CRS, 2017). However, this is not the only case. Eckstein (2011) points out that approximately 20 binational aquifers are the only relevant domestic water supply sources for many twin border towns: Puerto Palomas, Chihuahua, and Columbus, New Mexico, Naco, Sonora, and Naco/Bisbee, Arizona; Nogales, Sonora, and Nogales, Arizona; Sonoyta, Sonora, and Lukeville, Arizona; and Tecate, Baja California, and Tecate, California.

In general, the overexploitation of aquifers in the border region creates problems such as land subsidence, which has damaged housing and urban infrastructure. This is a problem in the El Paso, Texas/Ciudad Juárez, Chihuahua area (CRS, 2017). Groundwater reservoirs in other areas have been damaged or decreased in volume, generating a significant water deficit; this phenomenon has been repeated in other border areas, such as Tijuana, Baja California/San Diego, California and Nogales, Arizona/Nogales, Sonora, as well as in the Monterrey, Nuevo Leon metropolitan area (El Colef, 2019).

Actions in the United States have also yielded negative consequences for groundwater recharge in Mexico, as in the case of the All-American Canal, which brings water from the Colorado River. The United States lined sections of the canal, in an attempt to minimize water losses as the water traveled through it. However, this decision came at the expense of reduced groundwater recharge from the canal into shared aquifers beneath the canal, harming critical wetlands habitat and reducing water available for irrigation (Maganda, 2005; Scott et al., 2014). As a result, Mexico filed a case in an international court over lost water resources.

Changes in Climate and Water Availability

Climate projections by the Intergovernmental Panel on Climate Change anticipate that temperatures across the border region (more specifically, in the western monsoon climate region) will increase by as much as 2 to 4 degrees C by 2050 and 3 to 5 degrees C by 2100, coupled with decreases in precipitation of 5 to 8 percent (Wilder et al., 2010). Projected increased temperatures and drier conditions will exacerbate existing water stress and water quality issues on either side of the border (Wilder et al., 2013). Climate change is also exacerbating the declining quality and overall depletion of aquifers because the decrease in surface water caused by warming is both increasing the demand for groundwater resources and reducing the recharge rate (Wilder et al., 2010). Other ongoing stressors, including population growth, urbanization and industrialization, polluted water resources, and existing competition among water users, will compound these climate change impacts, making management across the border region more complicated (NASEM, 2018).

URBANIZATION AND INFRASTRUCTURE

The border region has experienced a growing economy based on various commercial activities—agricultural, mining, industrial, and services—many of which rely on using arid areas for tasks such as the production of fodder, mining, timber production, animal husbandry, and camping, among others (NASEM, 2018). The implementation in 1994 of NAFTA drove a great deal of population growth, sprawling urbanization, and industrialization, mostly in Mexico, as agriculture and industry shifted south from the United States. Exacerbated by huge consumer demand from the United States and a comparatively weak Mexican economy, this development has put tremendous strain on the region, riddled by poverty and natural resource constraints (Varady and Ward, 2009). It also pushed prices for agricultural commodities down, which disproportionately harmed poor farmers working unirrigated, communal land (Shah et al., 2004).

Urban development in Northern Mexico has been accelerated by the growth of the maquiladora[6] industry; for decades, the border region has seen population and growth rates above the national average (Peña Muñoz, 2018). One consequence of the presence of maquiladoras is that the traditional links between local production and consumption have been weakened or broken in several border towns (Díaz, 2009). The growth of the maquiladora industry in Mexico was established and maintained through the supply of low-wage labor and gender-based wages (Huesca et al., 2019). The development of Ciudad Juárez represents one of the most visible examples of this model (Peña Muñoz, 2018; Solís and Ávalos, 2017).

These economic drivers have resulted in the formation of informal, uninsured housing along border regions that is vulnerable to health and safety problems (Wilder et al., 2013). Worsening economic conditions for farmers have also spawned a rural-to-urban migration that has caused sprawling urbanization and the development of slum-like communities, resulting in poor environmental and public health outcomes (Spring, 2016). These health and environmental vulnerabilities are exacerbated by the fact that urban regions are expanding into areas that are prone to drought, wildfires, and flooding. This pattern of sprawl, as well as deficits in urban infrastructure, make the areas more prone to becoming urban heat islands—metropolitan areas that are warmer than the areas surrounding them (Wilder et al., 2013).

On the U.S. side of the border, communities referred to as "colonias" were granted official designation from the U.S. government in the 1990s. On both sides of the border, these often unincorporated and underfunded communities in both countries deal with complex and coupled challenges,

---

[6]The industry comprises factories in Mexico run by a foreign company whose products to that company are largely duty and tariff free.

such as the booms and busts of the agriculture and livestock industries (Hruska, 2019), violence and the drug war, the militarization of the border, inadequate living conditions, and low employment rates. Collectively, these challenges have depressed economic opportunities and social mobility and have exacerbated issues such as environmental pollution and access to safe and clean water supplies (Talmage et al., 2019). Despite the shared safety and economic concerns, U.S. colonias typically have more political autonomy, planned land use protocols, and basic infrastructure and services available to their residents than do the Mexican colonias (U.S. Department of Housing and Urban Development [HUD], 2020).

Poverty is widespread on both sides of the border, in poor U.S. counties, and unplanned Mexican colonias (Wilder et al., 2010). Per capita income among people living in the U.S. border counties is 85 percent of the average U.S. per capita income. If these counties were aggregated into a single state, it would be ranked 2nd highest in the nation in tuberculosis cases, 5th highest in unemployment, 39th in per capita income, 50th in health insurance coverage, and 50th in high school graduation rates (Soden, 2006).

Urban Water Infrastructure

Water and Wastewater Systems

Access to safe drinking water and sanitation services is inadequate for communities on both sides of the border, particularly in poorer communities with limited governmental and financial resources (Jepson, 2014). Population growth in both countries has outpaced the development of infrastructure in many urbanizing communities, adding pressure to the challenge of protecting public health.

In Mexico as a whole, 57 percent of the population lacked access to safely managed drinking water services[7] and 50 percent lacked access to safely managed sanitation services overall in 2017.[8] The colonias are particularly susceptible to water insecurity, as their populations are generally poor, marginalized, and often lack the critical infrastructure to deliver reliable water and sanitation services (Schur, 2017). While access to safe, potable water sources has improved over time, progress has been slow, and inadequate access to both safe water and sanitation remains critical (Wilder et al., 2013). Populations lacking piped water infrastructure are likely to be dependent on shared and often overexploited groundwater resources,

---

[7] More information is available at: https://data.worldbank.org/indicator/SH.H2O.SMDW.ZS?locations=MX.

[8] More information is available at: https://data.worldbank.org/indicator/SH.STA.SMSS.ZS?locations=MX.

which are generally not well-governed or protected by international treaties (Sanchez and Eckstein, 2020). Households that have hookups to water service might still encounter service disruptions, and water is not generally of drinking quality (Wilder et al., 2010). Water rationing has occurred in major cities in Mexico, including Hermosillo and Nogales in Sonora, and informal communities often rely on water from trucks (Wilder et al., 2013).

Poor water quality is a major challenge to the provision of safe drinking water along the border. Arsenic and fluoride water contamination, two inorganic contaminants associated with serious health problems (e.g., cancer, heart disease), are naturally present in the border area's groundwater aquifers, and concentrations of these contaminants have risen as a result of the over-pumping of aquifers, climate change, and rapid urbanization (Armienta and Segovia, 2008; Shaji et al., 2020). Drought and flooding can place even more pressure on a community's access to drinking water when traditional sources become contaminated or unavailable during such events and access to nontraditional sources, such as water trucks, might be unavailable, difficult to access, or cost-prohibitive (Wilder et al., 2010).

A recent study investigated water insecurity by looking at the very large 13,313 square kilometer-wide Mimbres Basin Aquifer, an arsenic- and fluoride-contaminated groundwater resource spanning southwestern New Mexico and northern Chihuahua (Schur, 2017). This region is home to underserved populations on both sides of the border. There are challenges and trade-offs associated with addressing drinking water insecurity in both nations. For example, centralized and decentralized reverse osmosis systems were implemented in the adjacent communities of Columbus, New Mexico, and Palomas, Chihuahua, respectively, to relieve drinking water contamination that has plagued both regions for decades (Schur, 2017). While implementing these systems addressed the poor water quality, it created new problems for water affordability because of the large energy and economic costs of the treatment systems. Household water costs in Columbus rose nearly 60 percent between 2008 and 2016 since all the water delivered to utility customers there is a blend of reverse-osmosis filtered and unfiltered groundwater. Thus, although a reliable source of piped water has been created, 70 percent of the population surveyed in 2016 considered the price charged for the water unfair (Schur, 2017). However, reducing these costs is difficult since the population of customers is small and the utility faces financial difficulties in generating adequate revenues to cover its costs. In Palomas, water piped to residents is still contaminated, but three stations distributed around the city dispense reverse-osmosis treated water for purchase. Nearly half of the residents (43%) surveyed reported that accessing these decentralized stations is difficult, and 18 percent of the population in 2016 still depended on contaminated tap water for drinking (Schur, 2017). These examples underscore the complexity of providing safe water services to poverty-stricken populations on either side of the border, particularly given the disparities in institutional support and funding mechanisms between the two countries (Giner et al., 2019; Schur, 2017).

High-salinity water, sewage flows, and flows of other polluted urban runoff present water quality challenges to be managed (Barker et al., 2000). Some of these water quality issues associated with the delivery of surface water from one side of the border to the other are managed through minutes to the 1944 U.S.–Mexico Water Treaty (Sanchez and Eckstein, 2020). The 1983 Mexico–U.S. Agreement on Cooperation for the Protection and Improvement of the Environment in the Border Area (i.e., the La Paz Agreement) was an important binational initiative to reduce and prevent pollution in the border region and provided a foundation for international collaborations that followed, which include NAFTA, BECC, and the Commission for Environmental Cooperation (Giner et al., 2019). Since the initiation of BECC (which has now been merged with NADB), the flows of untreated wastewater into shared water bodies have been an issue of great concern in the binational collaborations facilitated by NAFTA. Between 1994 and 2017, funding has been directed toward 59 wastewater treatment plants with a collective treatment ability of 450 million gallons a day serving more than 8 million residents in the United States and Mexico (Giner et al., 2019). These improvements to infrastructure and a marked reduction in sewage releases benefited an estimated population of 8.5 million residents during that period (Giner et al., 2019).

Despite these binational initiatives, wastewater treatment infrastructure is still woefully inadequate, particularly where burgeoning urbanization, industrialization, and population growth have boomed since the initiation of NAFTA. In Sonora, wastewater infrastructure statewide serves less than 40 percent of its population, and large populations of people live in colonias that are off the grid and might lack municipal services altogether. Tensions between the United States and Mexico have arisen over the costs of releasing treated wastewater flows originating from Nogales, Arizona, and Nogales, Sonora. Although the Nogales International Wastewater Treatment Plant treats the majority of wastewater, storm events still result in increased polluted flows from Mexico into the United States (Albrecht et al., 2018). Over time wastewater treatment capacity benefiting both sides of the border has increased due to binational cooperation of the United States and Mexico through BECC and the NADB (Giner et al., 2019).

Stormwater Management

Accelerating urbanization has added pressure to flood, stormwater, and wastewater management due to the expansion of impervious surfaces that impede the percolation of water back into soil and groundwater aquifers.

Average stormwater runoff can increase as much as 45 percent due to these decreases in infiltration, resulting in severe damage to private and public property through flooding and pollution flows, which can contain fecal matter, solid waste, oil, and sediment (Giner et al., 2019). Flood management is hindered by inadequate infrastructure in many parts of the border region, such as in Nogales, Arizona, and Nogales, Sonora. For example, in 2008, the U.S. Department of Homeland Security extended a portion of the border wall without coordinating with Mexican authorities, which resulted in catastrophic flooding of the Mexican side of the water near Nogales, Sonora, following a storm, when runoff that would have otherwise flowed northward (Wilder et al., 2010).

Despite some similarities in marginalized communities on either side of the border, the potential for outside resources is very mismatched. In the United States, unincorporated colonias can seek funding for infrastructure projects from a diversity of state and federal resources, in addition to international funding through the Border Environment Cooperation Commission. On the Mexican side, resources are much more limited and are typically only available from the federal government and binational agreements.

The costs accrued by emergency management departments and agencies in the United States, as well as by Mexico's natural disaster management agency, Fondo de Desastres Naturales, have grown markedly over the past few decades. Nevertheless, there is still no strong, coordinated, binational strategy to deal with stormwater flows (Giner et al., 2019).

Desalination

Desalination has been proposed as a solution to manage some water quality impairment of some border-region rivers, notably at the Yuma Desalting Plant in Arizona just upstream of where the Colorado River forms the border. While the Yuma Desalting Plant was built to reduce the salinity of water delivered from the United States into Mexico, this facility has rarely operated due to high operational costs and surplus flows of the Colorado River since its completion in 1992.[9] In 2012, the International Boundary and Water Commission entered an agreement to explore the feasibility of binational desalination for two prospective seawater desalination sites in Rosarito, Baja California, and Puerto Peñasco, Sonora, on the Sea of Cortes, which would export water to San Diego and Arizona, respectively (Wilder et al., 2016). There have been several proposals on how the United States and Mexico might share costs under a binational desalination regime. One proposed option would be for the United States to invest in a desalination facility in exchange for some portion of Mexico's water rights on

---

[9] More information is available at: https://fas.org/sgp/crs/row/R43312.pdf.

the Colorado River (Albrecht et al., 2018). However, to date, CONAGUA has not agreed to such terms, and many criticize existing proposals as one-sided, asserting that Mexico would incur the majority of the costs and environmental damages associated with the proposed binational desalination schemes, while the United States would reap a disproportionate share of the benefits (Albrecht et al., 2018).

Desalination has been touted as a "drought-proof" approach to supplying water, increasing the volumes of high-quality water and thereby improving water security and protecting water quality (Wilder et al., 2016). However, desalination comes with large environmental, economic, and social tradeoffs that have thwarted efforts to execute plans for building desalination facilities. From an environmental perspective, desalination by reverse osmosis, currently the most economical desalination technology, is energy-intensive, and therefore emissions-intensive when the electricity is generated from fossil fuel sources (King et al., 2013). The intake screens in these plants can harm marine ecosystems (which in turn could negatively affect tourism activities, in addition to the environment), and the necessary transportation infrastructure, such as pipes, would need to cross delicate ecosystems (Albrecht et al., 2018). In addition to generating a potable water stream, desalination also generates a brine stream that is difficult to dispose of in an environmentally benign way. In part because of its high energy costs, desalination is also costly, driving up the cost of water compared to standard surface water supplies, which could negatively affect impoverished communities on either side of the border (Wilder et al., 2016).

AGRICULTURE AND LIVESTOCK

Since the beginning of the 20th century, Mexico's proximity to the U.S. market has driven the deployment of irrigation infrastructure in the deltas and valleys of the Colorado, Sonora, and Rio Grande/Bravo rivers, and this has sustained commercial agriculture in the region. Therefore, it is not surprising that this region has also been the historical scene of agricultural disputes over the control of these means of production.

Despite the historical existence of cross-border agricultural systems, particularly in the western border region,[10] agricultural activity in border towns still exhibits dramatic contrasts on both sides of the border.

According to statistical data from Mexico and the United States (Sistema de Información Agroalimentaria y Pesquera [SIAP], 2018a; U.S. Department of Agriculture [USDA], 2017), industrial agriculture prevailing on both sides

---

[10]There are documented links between the agricultural capital of Sonora and Arizona (Pavlakovich-Kochi, 2006), and the transfer of U.S. companies in shaping the agricultural scenario of the Mexican side of the border.

of the border is aimed directly at the United States. Most of the products are for food chains, but in the case of cotton, it is for industrial purposes.

The relevance of agricultural activity in the region to the economy of each country is quite different. While in Mexico agriculture represents one of the main areas of commerce, by way of export products, in the United States agriculture is secondary to livestock production, except for some irrigated valleys in California, Arizona, and Texas. According to the 2017 U.S. Census of Agriculture, concentrated livestock activity in Texas, California, and Arizona continues to focus on sheep and lamb production, although there is also significant cattle production along the border. A higher concentration of cattle is found in the far south between California, Arizona, and Texas, with milk production in areas near Phoenix, Arizona, and meat production found primarily in Texas (USDA, 2017).

On the Mexican side, agriculture, more than livestock, drives the configuration of the region. Agricultural production in the six border states accounts for 22.6 percent of Mexico's domestic agricultural production, a proportion of 32.5 percent relative to the value of domestic production in irrigation mode, compared to just 5 percent of seasonal domestic production. In addition, Chihuahua and Sonora concentrate two-thirds of the border states' production value and, in both cases, it is irrigation production that supports this dynamism. Focusing only on export agriculture, the border states make an even greater contribution, making up 75 percent of domestic production and comprising 65 percent of the total area in the country devoted to export culture. This concentration is centralized in Baja California and Sonora; these two Mexican states comprise 64.6 percent of exports, and 74.3 percent of its value (SIAP, 2018b).

Livestock

Livestock production in the Mexican border states accounted for 15.7 percent of the country's livestock production and is focused on meat production, followed by the marketing of live cattle, goats, and pigs (Hernández Pérez, 2019; SIAP, 2019). The state of Sonora concentrates the livestock production of the region, with meat products, live cattle, and eggs. And while not all this production occurs on the border margin, there is a significant concentration of farms and ranches in the Sonora municipalities of Hermosillo, Navojoa, and Cajeme.

Agricultural Water Use and Environmental Effects

Water use for agriculture and ranching accounts for approximately 80 percent of total water usage across the shared border region (NASEM, 2018), dominating consumptive water use on either side of the border—although rapid urbanization and industrialization are increasing demand for non-agriculture water uses (Wilder et al., 2010). In Arizona, agriculture represents 70 percent of consumptive water use, while in Sonora, Mexico, it represents 86 percent (Wilder et al., 2010).

Agro-industrial production in the United States is centered in four areas of intensive irrigation. Two of the irrigation centers are located near the banks of the Colorado and Gila rivers at the California-Arizona border; their main outputs are fruits and vegetables. A third region, just south of New Mexico in the valley formed by the Rio Grande at Las Cruces, produces alfalfa and pecans. The fourth irrigation center in Hidalgo, Texas, running to the mouth of the Rio Grande/Bravo, primarily produces cotton. As a consequence of heavy land use, the depletion of water has been recognized as a problem along the entire U.S. southern border, one characterized by low soil productivity, particularly in Southern California, Arizona, New Mexico, and the western tip of Texas (Miller et al., 2012). It is therefore not surprising that this region has also been the historical scene of agricultural disputes over the control of these means of production.

Recent changes in agricultural land use on either side of the U.S.–Mexico border has affected water usage. In the United States, more water is withdrawn for irrigated cropland than for urban land, on average. In Mexico, this trend is reversed, because urban regions generally have higher population density, despite lower per capita usage. Thus, a parcel of irrigated cropland converted to urban use in the United States would yield net decreases in water use, while in Mexico such a conversion would yield a net increase in use (Bohn et al., 2018).

NAFTA drove an expansion in irrigated agriculture in Mexico in the 1990s, particularly for the cultivation of fruits and vegetables for export, the overwhelming majority of which are exported to the United States. Despite the environmental provisions in the trade agreement, this expansion in agricultural production has resulted in the overdraft and salinification of the region's groundwater aquifers, because it has put more pressure on limited surface water resources. Furthermore, agricultural production in the United States decreased during the same period, as markets shifted to the Mexican side, which brought agricultural water usage reductions to the former at the expense of increasing the water usage for land south of the border (Bohn et al., 2018). As a consequence of heavy land use, the depletion of water has been recognized as a problem along the entire U.S. side of the southern border, one characterized by low soil productivity, particularly in Southern California, Arizona, New Mexico, and the western tip of Texas (Miller et al., 2012).

The dynamic growth of agro-industrial production, the intensification of resource exploitation, and the regulatory heterogeneity under which the region's agro-industrial systems operate have triggered a variety of environmental and social problems beyond just water exploitation concerns. A review of the main cross-border management documents of the past few decades has shown the following primary problems:

- the practice of agricultural burning as a contributing factor to air pollution (EPA-SEMARNAT, 2012, p. 18);
- the intensive use of pesticides and the lack of regulation and education regarding their application, storage, application, and ultimate disposal (U.S. Environmental Protection Agency and Secretaría de Desarrollo Urbano y Ecología [EPA-SEDUE], 1991, pp. III-38–III-39; EPA-SEMARNAT, 2012, p. 29);
- contamination of groundwater and bodies of water because of pollution derived from the intensive use of pesticides and fertilizers (EPA-SEMARNAT, 2012, pp. 11, 15);
- water depletion and/or distribution exacerbated by high-water demand activities, such as industry and agriculture, occurring in an arid climate (EPA-SEMARNAT, 2012, p. 5, 19–20); and
- the marginalized working and living conditions of temporary agricultural workers, particularly their negative effect on workers' health and education (EPA-SEMARNAT, 2012, p. 6; Osuchukwu et al., 2017; Villarejo, 2002).

Compounding the other causes of environmental degradation in the border strip is poor management of agricultural runoff (and wastewater), a problem that has been documented by the GNEB since its creation in 1992 (GNEB, 2014). Water quality issues spurred by these polluted streams of agricultural runoff have been so severe that they have generated "dead zones" at the mouths of border rivers, affecting populations of aquatic species and the people who depend on them.

One case that illustrates the coupled environmental impacts that can be exacerbated by intensive agricultural operations is the New River (Río Nuevo), which drains in the Salton Sea in California. It receives agricultural and urban runoff with a high concentration of pesticides, untreated wastewater, and industrial waste. This river is considered one of the most polluted in North America, and although binational cooperation has reduced its pollution levels, it remains a major problem for the population of the Imperial Valley (California Environmental Protection Agency [CalEPA], 2020). In particular, the border area presents serious problems not only of water pollution but also of air pollution, which was aggravated by the commissioning of fossil fuel-based power plants at the beginning of the 21st century (Ramos and Reyes, 2006). Data collected in 2016 by the World Health Organization found that the city of Mexicali, had some of the highest average PM10 levels[11] in North America and the sixth-largest in the entire continent, with 85 micrograms per cubic meter (James, 2019). Between 2010 and 2016, at least 78 people died of asthma and 903 more people from chronic obstructive pulmonary disease in that city. Overall, this contamination is estimated to cause around 300 premature deaths annually in Mexicali (James, 2019). Mexicali's is not the only case of air pollution on the border: in 2017, the EPA noted that the number of days in which air pollution reached a level classified as a risk to vulnerable people were 22 for El Paso, Texas; 27 for Las Cruces, New Mexico; 33 for El Centro, California; and 55 for San Diego, California (Eades, 2018).

In the 1990s, mirror cities shared similar air pollution levels, but they became increasingly dissimilar over the years due to differences in regulatory standards between the two countries. According to the Commission for Environmental Cooperation (CEC, 2004), this was the result of less stringent air quality standards implemented in Mexico, where those standards are perceived as objectives rather than as requirements to be implemented (see also Cresswell et al., 2009).

Agricultural burning, pesticides, and water scarcity have been addressed repeatedly in the binational programs agendas coordinated by the EPA and SEMARNAT, Mexico's environmental agency. Work environment issues for agricultural day laborers were added to this agenda as part of a broader perspective that includes the deterioration of workers' health resulting from pesticide use. While the contamination of groundwater by agricultural pollutants is regarded as a transcendent binational concern, there is no clear or specific work agenda around this in either country.

Financial Tensions

Natural resource challenges have been coupled with recent agricultural reforms in Mexico that have disproportionally affected poor farmers. Much of the agricultural development and intensification that occurred in the 20th century in the arid north was driven by large-scale irrigation projects executed by the government in "underutilized land." The government granted direct agricultural subsidy payments enabling the development of *ejidos*, which are non-sellable communal land-use rights, mostly established for livestock and crop production (Hruska, 2019; Shah et al., 2004). The Mexico border region has approximately

---

[11] PM is particulate matter, also called particle pollution. PM10 are inhalable particles that typically have diameters less than 10 micometers. See: https://www.epa.gov/pm-pollution/particulate-matter-pm-basics for additional discussion.

2,336 *núcleos agrarios*[12] (agricultural units) (Registro Agrario Nacional, 2019), many of which are large state-supported ejidos with large irrigation areas, created mostly in the 1930s and 1940s.[13] In the north, ejidos have helped to generate incomes in rural areas, but in general, they have failed to produce sustainable profits for their members (ejidatarios) due to small parcel sizes, poor land quality, and initial poverty, which have precluded the efficient development of lands.

Since the 1990s, the development of NAFTA, and the consequent amendment of Article 27 of the Mexican Constitution, there has been a push for the privatization of ejidos and water and electricity systems (Shah et al., 2004). In 1992, Article 27 was modified to allow the ejidatarios, to mortgage or sell their land, as long as they had at least two-thirds support from their members (Shah et al., 2004). This reform resulted in the transfer of public land to private, wealthier entities, making it even more difficult for other poor ejidatarios to compete (Hruska, 2020). As a result of these reforms, governmental agricultural buying programs and subsidies were markedly reduced or eliminated, increasing the cost of production for rural farmers, and many ejidatarios were no longer able to afford necessary farming inputs. Exacerbating these financial tensions was a decade-long drought that made it nearly impossible for farmers dependent on rain-fed agriculture and rain-fed forage for their livestock. Drilling new groundwater wells is cost-prohibitive for a vast majority of ejidatarios. At the same time, crop prices were generally pushed down, prompting the emigration of many farmers to the United States (Vásquez-León et al., 2002).

Thus, areas where the majority of poor farmers and cattle owners share these communal lands are particularly vulnerable to water scarcity (Hruska, 2020). Although NAFTA expanded some markets for high-value crops like chiles and onions as crop production shifted from the United States to Mexico, declining governmental support and prolonged drought over the past few decades have made it difficult for many farmers to survive (Vásquez-León et al., 2002). In short, agriculture has continued to expand in northern Mexico since the implementation of reforms, but the massive reductions in government resources that propped up poor farmers who would not otherwise be self-sustainable created more economic stratification between farmers, that is, between those with the resources to survive with their financial resources and those who could no longer make a livable income (Hruska, 2020).

---

[12] This number was generated by applying a 100-kilometer buffer south of the border to the most recent available database from Registro Agrario Nacional, using the online GIS. A freely downloadable dataset from 2017 is available at: http://132.248.14.102/layers/CapaBase:ran_nacional_2017_wgs84, but using this older dataset yields a different number (2,088) of núcleos agrarios.

[13] More information is available at: https://mexico.leyderecho.org/nucleo-agrario/.

MINERAL RESOURCES AND MINING

Mining is an activity that has played a very important role in the region, historically, in economic development (representing about 10% of the country's Gross Domestic Product [GDP]) (Aguilar-Pesantes et al., 2021), as well as the location of human settlements and urban growth. However, it is also an activity that has impacted the environment, as it is a consumer of water and energy. In addition, copper, more so than silver, plays an important role in the economy on both sides of the border. Mexico is the world's largest silver producer, representing greater than 23 percent of global production in 2019.[14] Chihuahua and Sonora are large silver-producing regions, representing 20.8 percent and 10.8 percent of the country's production, respectively. Mexico also is the ninth-largest producer of the world's gold,[15] with Sonora and Chihuahua representing 33.2 percent and 17.2 percent of national gold production, respectively.

Sonora is Mexico's largest mining region with gold, silver, copper, and molybdenum mining in its more mountainous regions. The state's mining sector, representing about 17 percent of the state's GDP, provides clear examples of the tensions between economic productivity and environmental sustainability in the region, given ongoing concerns of water scarcity and climate change (Zuniga-Teran et al., 2020). The state is home to the Buenavista del Cobre (Cananea) mine, the fourth largest copper mine in the world,[16] located 35 kilometers south of the U.S.–Mexico border (Mendoza-Lagunas et al., 2019).

The mining industry has placed increasing priority on environmental sustainability and social responsibility. Many companies are developing and implementing active and ongoing programs with a focus on community relations, creating social value where they operate alongside an emphasis on environmental performance. A small number of high-profile mining accidents, such as the one in 2014 at the Buenavista del Cobre mine in Cananea, Sonora, 22 miles south of the U.S.–Mexico border (Mendoza-Lagunas et al., 2019), continue to serve as a reminder of the importance of advancing technology and governance for human and environmental safety. Partnerships are emerging among leading U.S. universities and mining companies that are making efforts to involve Mexican counterparts to share the benefits of research and further cooperation. For instance, the University

---

[14] More information is available at: https://www.statista.com/statistics/253339/leading-silver-producing-countries/.

[15] More information is available at: https://www.gold.org/goldhub/data/historical-mine-production.

[16] More information is available at: https://www.mining-technology.com/features/feature-the-10-biggest-copper-mines-in-the-world/.

of Arizona's Lowell Institute for Mineral Resources[17] works to advance responsible mining through binational and global networks to educate stakeholders on innovations in modern mining technology and practices.

Similar to the observation above for renewable energy, mining holds major potential for binational economic development and trade, and, therefore, for opportunities for binational sustainability partnerships.

Mining companies that are active on both sides of the border include Grupo México (in Arizona and Sonora) and Capstone (in Arizona and Zacatecas). Companies operating in one country alone are becoming increasingly aware of the benefits of binational partnerships that provide access to knowledge, resources, training, and technology for sustainability in arid regions through collaboration with universities whose research often transcends borders. Companies exploring these opportunities to date include ASARCO and Fresnillo plc.

CONVENTIONAL AND RENEWABLE ENERGY

The United States and Mexico are both large energy-producing and energy-consuming countries. The border region itself is a source of both renewable and nonrenewable energy resources, and it is the focal point of an important articulated network for the flow of energy between the two countries. This network has densified over the past two decades as both countries and Canada move toward a North American energy integration (U.S. Government Accountability Office [GAO], 2018). In terms of partnerships, the abundant renewable energy resources in the binational region present large opportunities for collaboration between the two countries.

Binational Energy Characteristics and Asymmetries

In 2018, the United States and Mexico represented the 1st and 11th largest petroleum-producing countries, respectively, and the 1st and 27th biggest natural-gas producing countries in the world, respectively (U.S. Energy Information Agency [EIA], 2020a). In terms of total renewable energy generation (nuclear, renewable, and other production), they represented the 1st (United States) and 20th (Mexico) largest generators in the world. However, the two countries differ markedly in consumption. In 2018, the primary energy consumed in the United States amounted to 310 million British thermal unit (BTU) per person (ranking 10th in the world), whereas in Mexico it was 63 million BTU per person (88th in the world), reflecting wide disparities in the social and economic fabrics of the two countries (EIA, 2020a).

---

[17] More information is available at: https://minerals.arizona.edu/.

The U.S. total supply of primary energy in 2019, that is, its domestic energy production plus energy imports from other countries, minus exports and international bunkers amounted to 2,200,788 ktoe (kilotonne of oil equivalent) with petroleum, natural gas, and coal representing 36 percent, 33 percent, and 13 percent, respectively, while nuclear (9.9%), hydropower (1.1%), biofuels and waste (4.9%), and wind and solar (2.1%) made up the balance of the nonfossil fuel supply (International Energy Agency [IEA], 2020a). That same year, fossil fuels represented 80 percent of total U.S. primary energy consumption (petroleum: 37%, natural gas: 32%, and coal: 11%). Nuclear and renewable energy represented 8 percent and 11 percent of primary energy consumption, respectively (EIA, 2020b).

Over the past two decades, the energy landscape on the U.S. side of the border region has shifted dramatically chiefly due to three factors: the growth of renewable energy across the border states, the rapid growth of the fracking industry, and the decline of the coal industry (Peer and Sanders, 2018). Texas generated over a quarter of the country's wind power and produced approximately a quarter of its natural gas in 2019 (EIA, 2020c). California and Arizona represent the first and second-largest producing U.S. states in terms of solar photovoltaic generation. Even New Mexico has grown in energy output from being the seventh-largest producer of crude oil in 2013 to the third-largest by 2018, due to growth in the shale oil industry, while also generating about a fifth of its electricity from wind that year (EIA, 2020c). The fossil fuel and renewable energy resources in these states have enjoyed extraordinary growth due to technological advancements, which have driven down the price of energy production. Mexico shares many of the same types of geologic and renewable resource potentials as these U.S. states.

In 2019, Mexico's total primary energy supply, including imports, amounted to 184,021 ktoe of energy, comprised of 45 percent petroleum, 38 percent natural gas, 5 percent biofuels and waste, 1.6 percent nuclear, 1.4 percent hydropower, and 2.8 percent of wind and solar renewables (IEA, 2020b). The country's total primary energy supply grew by approximately 46 percent between 1990 and 2018. In 2019, petroleum, natural gas, and coal represented 43 percent, 42 percent, and 7 percent of domestic total primary energy consumption, respectively, while nuclear, hydroelectricity and non-hydropower renewables represented 1 percent, 3 percent, and 4 percent, respectively (EIA, 2020d). In 2019, Mexico was the second-largest import source for the United States, importing 1.2 million barrels per day of petroleum (EIA, 2020e).

Mexico's energy sector has been in decline over the past few decades, which spurred a series of institutional reforms in 2013 that have influenced the path of its energy industry (Vietor and Sheldahl-Thomason, 2017). The reforms were adopted in part because of very low oil prices, which put huge financial pressure on Mexico's state-owned petroleum company, Pemex, which has historically controlled the vast majority of oil and gas development in the country (Weijermars et al., 2017). Under the control of Pemex, oil production peaked in 2004 and has declined to nearly half of that peak in the 15 years since (Gross, 2019). The reforms ended Pemex's 75-year monopoly in the oil and gas industry (Vietor and Sheldahl-Thomason, 2017). Similarly, the electricity sector, also controlled by a state-owned monopoly (Comisión Federal de Electricidad [CFE]) was challenged by an aging infrastructure and high prices before the reforms. Mexico's former president, Enrique Peña Nieto, implemented reforms to address these declining industries with two goals: to create competition by attracting new technologies and market participants and to bring in capital to ensure the resources to meet growing energy demand (Gross, 2019; Vietor and Sheldahl-Thomason, 2017).

Renewable energy development accelerated after the energy reforms as international companies were incentivized to invest and operate in the country (Gross, 2019; Vietor and Sheldahl-Thomason, 2017). The total resource potential, based on very strong solar, wind, and geothermal resources, is large enough for Mexico to be a global leader in renewable energy development (Sanders et al., 2013).

Energy Transitions and Environmental Impact

Energy transitions, particularly those associated with the burgeoning U.S. shale industry, have spurred environmental concerns. The Eagle Ford shale region of southwest Texas has seen large surges in oil and natural gas development. Production in these shales and other tight formations has become economical only in the past two decades due to advancements in the coupling of hydraulic fracturing and horizontal drilling techniques that make it possible to extract fuels from impermeable rock (Jackson et al., 2013; Vidic et al., 2013). This development has come with a lot of environmental and social tradeoffs. Developing a well typically requires 10,000 to 30,000 cubic meters of water, depending on geology and the production methods (Rahm and Riha, 2014). It is estimated that 80 percent of the water used for hydraulic fracturing is freshwater and that 90 percent of this freshwater is sourced from groundwater (Mohtar et al., 2019).

The growth of the hydraulic fracturing industry has also triggered water quality concerns, such as gas migration into groundwater aquifers, accidental spills of toxic fracturing fluids, and the safe handling of wastewater produced during production (Vidic et al., 2013). There are other, non-water impacts as well, such as land degradation, air pollution, and increased greenhouse gases (Mohtar et al., 2019). There has also been a growing financial burden placed on communities that have had to pay for much of the damage caused to their communities by increases in hydraulic fracturing that resulted in increases in truck traffic for water management (and the associated increases in road degradation and road accidents), as well as public health consequences (Jackson et al., 2013; Mohtar et al., 2019). Although energy, and particularly renewables, represent a major potential source of economic development, cross-border trade, and binational sustainability initiatives, efforts by the study committee to invite energy-sector representatives to the stakeholder workshop were unsuccessful; thus, this topic is absent from the report. Instead, the committee opted to list this as a theme for future partnership efforts.

RESOURCE GOVERNANCE, INNOVATION, AND PARTNERSHIPS

Water Governance

In comparison with other water-scarce countries sharing borders and waterways, the relationship between Mexico and the United States is unique (Bonner and Rozental, 2009). The 1983 Mexico–U.S. Agreement on Cooperation for the Protection and Improvement of the Environment in the Border Area (i.e., the La Paz Agreement) was an important binational initiative to reduce and prevent pollution in the border region and provided a foundation for international collaborations that followed, which include NAFTA, the Border Environment Commission, and CEC (Giner et al., 2019).

Water shortages on the Rio Grande and Colorado rivers have spurred many international water management disputes, but they have also motivated many successful instances of international collaborations (Sandoval-Solis et al., 2013; Wilder et al., 2010). Despite successes, managing shared water resources is incredibly difficult in arid regions, and stressors—such as climate change, the differences in the way that water is managed in each country, population growth, shifts in urbanization and industrialization patterns, and limited financial resources—will continue to add pressure to the management of shared surface water and groundwater resources.

In general, water management in the United States is more decentralized and fragmented than water management in Mexico (Carter et al., 2017; Gerlak, 2006). In the United States, water rights and water-related laws and governance are administered by a patchwork of agencies at the federal, state, and local levels, which represent a variety of priorities and environmental protections (e.g., water development, water quality, ecological flows, irrigation withdrawals, interstate, and international water sharing). Because the United States generally has more financial resources and more actors with a stake in local governance of water, it is generally easier for projects to be financed there than in Mexico, where there are far fewer options (Medgal and Scott, 2011).

In Mexico, water management is generally more centralized. It is controlled by the country's national water commission, CONAGUA, which operates based on the country's Law on National Waters. This centralization can make it hard to prioritize local projects, since the tax base supporting CONAGUA is national and, by design, focused on projects throughout the country (Goetz and Berga, 2006). While Mexico's centralized control of the water system can enable more streamlined decision making in comparison with the United States (Wilder et al., 2013), reform efforts in recent decades have sought to decentralize aspects of Mexico's water management regime to engage more local stakeholders and spur more participatory governance (Wilder and Romero, 2006). These reforms have produced positive outcomes, particularly in the creation of local watershed districts and irrigation districts in northern states. However, some argue that mechanisms to shift water management to local authorities have left many communities without adequate resources to run water utilities and fund infrastructure, exacerbating existing poverty, corruption, and issues related to transparency (Scott and Banister, 2008).

In addition to the challenge of managing surface water flows, there are a range of groundwater resource governance needs in the U.S.–Mexico border region. As discussed in the 2018 workshop, the overdraft and salinification of aquifers are major issues on both sides of the border (NASEM, 2018). Within each country, there are also asymmetries in groundwater ownership (Megdal and Scott, 2011). Groundwater in Texas, for example, is considered private property, while in Mexico it is national property (Sanchez and Eckstein, 2020). These asymmetries in institutional governance are considered by many to be a primary barrier to a binational treaty that would better manage the common pool resource (Albrecht et al., 2018; Mumme, 2005; Sanchez and Eckstein, 2020).

The 1944 U.S.–Mexico Water Treaty addresses shared surface waters and is largely silent on groundwater. However, several minutes to amend the treaty have been passed that address groundwater, among other issues, including Minutes 319 and 323 (Buono and Eckstein, 2014; Mumme, 2020). The United States tends to have more enforceable protections for groundwater overdraft than Mexico (although protections vary from state to state, as discussed below), and the latter has seen much more expansion of irrigation due to weak overdraft protections (NASEM, 2018).

In the United States, approaches to groundwater governance are uneven and span many levels of local, regional, and state governments. For example, Arizona's 1980 Groundwater Management Act, managed by the state's Department of Water Resources, was implemented to protect groundwater aquifers from overdraft through such provisions as prohibiting irrigated agriculture on new land, while Arizona's Department of Environmental

Quality enforces water quality standards.[18] Additionally, there are often more regional approaches to groundwater management, carried out by local governments (e.g., the Santa Cruz Active Management Area in the Santa Cruz basin of Arizona) (Scott et al., 2012). While there are no formal binational protections for groundwater, the U.S.–Mexico Transboundary Aquifer Assessment Act was designed to conduct and improve data sharing and scientific research on water quantity and quality issues across shared aquifers (Callegary et al., 2016). Subsequent binational negotiations between members of the International Boundary and Water Commission from the United States and Mexico led to the 2009 signing of the "Joint Report of the Principal Engineers Regarding the Joint Cooperative Process United States-Mexico for the Transboundary Aquifer Assessment Program," which provided a framework for the use and joint study of shared aquifers.[19]

Over time, variable and declining precipitation patterns, along with rising competition for water, have decreased the amount of surface water available for agriculture. As a result, the exploitation of groundwater aquifers has worsened over time. In 2003, a night-time tariff was introduced to promote more agricultural productivity; however, these tariffs have incentivized over-pumping, exacerbating depletion (Scott, 2013). Although these tariff programs have successfully transformed otherwise desert-like northern regions in Mexico into productive agricultural regions that produce large quantities of fruits and vegetables for export, the resulting levels of groundwater depletion have reduced the adaptive capacity of the region to respond to future water scarcity (Sietz et al., 2011).

Effects of Trade Policies on Natural Resources

The strong trade integration that NAFTA opened led to changes in the cross-border agricultural landscape, particularly in Mexico. The constitutional reforms carried out under the signing of NAFTA have had a strong impact on Mexico's entire primary sector, mainly due to the entry of tariff-free agricultural goods and the parallel elimination of the marketing and production support system implemented decades ago by the Mexican government. These processes led to the disintegration of much of the agricultural production supplied by the domestic market, which encouraged productive specialization in export goods such as beef, vegetables, and fruit. As a result, the western border region has specialized in the production of fresh "winter" fruits and vegetables that respond to the demand of the U.S. market and are also exported to European and Asian markets.

---

[18] More information is available at: https://new.azwater.gov/sites/default/files/media/Arizona%20Groundwater_Code_1.pdf.

[19] More information is available at: https://www.usgs.gov/centers/ot-water/science/transboundary-aquifer-assessment-program-taap?qt-science_center_objects=0#qt-science_center_objects.

This conversion has also promoted the industrialization of the sector, which has intensified the dispute over inputs such as water and land, as well as raising demand for a labor force mainly for vegetable and fruit growing. Although legislation introduced in 1992 opened the possibility that community-held ejido agricultural land could be sold, but only a small proportion (7%) has been privatized in the border region, mostly in industrial and suburban areas (Vidaurrázaga Obezo, 2003).

Asymmetries in communities on either side of the border affect farmers' resilience in managing their livelihoods in times of drought. Agriculture in the United States has been less vulnerable to shocks, such as drought, due to technological interventions, including more efficient irrigation. However, pumped groundwater for irrigation is typically more expensive than surface water deliveries and still tends to be a limiting factor as to whether or not farmers can continue to operate (Vásquez-León et al., 2002).

As a result of rising water costs, as well as threats of disruptions in productivity because of prolonged drought, many communities have adopted technology-centric methods of ensuring a stable water supply to irrigate crops through methods such as drip irrigation and center-pivot irrigation (Vásquez-León et al., 2002). However, even with water-efficient irrigation, groundwater pumping is expensive, which has led to interesting tradeoffs between water use and the economics of crop production (Vásquez-León et al., 2002). While the increases in the cost of irrigation in groundwater-dependent regions have led to decreases in the agricultural land cultivated, there has also been a shift to crops that produce more economic value to offset irrigation costs, and some of these crops have high water needs (Vásquez-León et al., 2002). As a result, efforts to reduce water usage have been undermined by the movement toward more water-thirsty crops in groundwater-dependent regions.

Changes in Energy and Climate Governance

Even though it is a major fossil fuel-producing country, Mexico has established itself as a leader in international climate negotiations with its decarbonization goals, particularly in comparison to other emerging economies (von Lüpke and Well, 2019). Mexico passed its General Law on Climate Change in 2012 under President Felipe Calderon, which established a goal of reducing greenhouse gas emissions by 50 percent below 2000 emissions levels by 2050. A few years later, Mexico introduced a target to reduce greenhouse gas emissions by 2030 by 22 percent, relative to a business-as-usual scenario.[20]

---

[20]More information is available at: https://www4.unfccc.int/sites/ndcstaging/PublishedDocuments/Mexico%20First/MEXICO%20IND C%2003.30.2015.pdf.

Since the election of President López Obrador in 2018, Mexico's leadership has deprioritized clean energy and climate change mitigation actions.

In March 2021, the Obrador administration approved a fast-tracked bill that modifies Mexico's Electric Industry Law and rolls back much of his predecessor's energy reform initiatives.[21] Ending Mexico's energy reforms has the potential to limit the domestic production of oil and gas resources in shale basins and difficult-to-access offshore locations that might require the expertise of international producers outside of Pemex (e.g., producers with expertise in developing U.S. shales) (Weijermars et al., 2017). Fracking has been a growing industry on the U.S. side, but despite its rich shale basins just south of some of Texas's very productive shales, fracking is still a very new industry to Mexico. Two major trends have created a favorable environment for the domestic shale gas industry in recent years, namely a growing dependency on natural gas imports and regulatory reforms in the oil and gas sector, opening the country to foreign producers.

In 2002, Mexico became a net importer of natural gas, much of which is imported in the form of expensive liquified natural gas, incentivizing methods to grow domestic production. Pemex commenced exploration of the Eagle Ford shale play (shared with Texas), just south of the U.S.–Mexico border, in 2010, but no gas was identified until 2013 (Weijermars et al., 2017). The real game-changer for the fracking industry came a year later through the 2014 energy sector reforms. These have made it easier for foreign operators to produce in the country, thereby bringing in the expertise needed to produce in more difficult shale regions. The reforms spurred significant energy investments, much of which have been directed toward the still-nascent fracking industry in the northern regions of the country (Gross, 2019). Thus, the pivot by President López Obrador to direct control back to Pemex might stall the continued development of these harder-to-access resources (Weijermars et al., 2017). Yet, Mexico contains the world's sixth-largest reserves of shale gas, concentrated in the north, where water-scarcity issues, particularly in terms of overexploited groundwater aquifers, are already pronounced (Weijermars et al., 2017). Thus, anticipated growth in the industry, which has been primarily concentrated in the United States at this point, will need to be undertaken with environmental protections in mind.

Energy development in the United States is much more market-based than in Mexico. The federal government does not have nearly as much power to influence the dynamics of the energy industry as it has in Mexico, as there are also state and local policies that can affect the development of energy policy. Thus, the pullback of energy reforms in Mexico in recent years is likely to have a much bigger potential to affect the energy landscape than any leadership change in the United States.

---

[21] More information is available at: https://www.elfinanciero.com.mx/nacional/senado-aprueba-en-lo-general-la-reforma-electrica-de-amlo.

SUMMARY

As described in Chapter 1, the May 2018 National Academies workshop "Advancing Sustainability of U.S.–Mexico Transboundary Drylands," applied a thematic lens to the region's challenges to explain sustainability dynamics more effectively than the conventional approach that looks at each resource and phenomenon individually. In considering partnerships that most effectively address binational sustainability, it is critical that these four themes in sustainability dynamics—interactions and flows; scarcity and abundance; shocks and stressors; and governance, innovation, and partnerships—be placed at the forefront.

Interactions and flows: The transborder region is constantly in flux with dynamic interactions and flows of people, resources, and ever-changing political arrangements. As a result of rapid industrialization and a decrease in agricultural production, many farmers have been forced to move to more urban regions, exacerbating the trend toward urban sprawl in unincorporated, slum-like communities. Changes in trade regimes have also shifted the region's demographics and social activities over time. The large flow of commodities and industrial products across the border is more than matched by the movement of people going to Mexico as recreational or medical tourists and to the United States seeking jobs or escaping hardship and violence in Mexico or other places. The flow of economic migrants and refugees into the United States occurs both legally and illegally, a reality that often dominates the political agenda in both countries. Illegal trafficking of drugs and firearms between the two countries is a continued threat to population health and safety.

Scarcity and abundance: The U.S.–Mexico border region embodies both scarcity and abundance—rich in ecological, natural resource, and mineral wealth, while also characterized by aridity and desertification. Water security remains one of the chief concerns for the region, particularly as industrialization, shifts in economic opportunities, migration, and the proliferation of large agricultural developments on both sides of the border have increased water strain in recent years. The U.S.–Mexico border region struggles with severe water stress, exacerbated by irrigation and overgrazing activities, deforestation, and severe soil degradation associated with agricultural production. These challenges are particularly prominent in Mexico's arid and semi-arid regions, which, despite receiving a small fraction of the country's total precipitation, have most of its irrigated land (Díaz-Caravantes and Wilder, 2014).

Shocks and stressors: Rapid population growth in and around "mirror cities" (i.e., urban regions situated adjacent to one another on either side of the border), whose expansions have been characterized by sprawling urbanization and the development of formal and informal communities, have created anthropogenic shocks and stressors in the region. The lack of adequate infrastructure for essential needs such as basic drinking water and sanitation services jeopardizes the public health and security of residents while straining natural resources, particularly in the context of climate change. Furthermore, differences between the United States and Mexico in terms of critical infrastructure, economic security, regulatory environment, culture, and language, often hinder efficient binational management of shared resources such as surface and groundwater. They also create challenges for mitigating shocks and stressors, such as excess flows of wastewater during flooding events.

Governance, innovation, and partnerships: The region is also a crucible for developing sustainability in governance, innovation, and partnerships. Many of the shifts in recent decades toward industrialization, urbanization, and migration were driven by the implementation of NAFTA in 1994, which resulted in vast increases in the international trade of agricultural commodities grown along the border and large decreases in the consumption of agricultural goods in the domestic market. Drivers of border challenges are often heavily determined by national policy initiatives. Differential financial resources and opportunities for public participation, as well as Mexico's comparatively recent shift to a plural party democratic process, result in asymmetries in the countries' capacities to respond to exogenous policy stressors affecting the border region.

The shift in Mexico from the local consumption of domestically produced goods to production for the transnational market led to the proliferation of maquiladoras/assembly plants and adversely affected farmers in the region by pushing down the price of agricultural commodities.

Additionally, while trade has been at the forefront of the U.S.–Mexico relationship, there have been successful partnerships in other areas. Both Mexican and U.S. communities along the border face common threats, such as water scarcity, inadequate infrastructure, and land and soil degradation, which have led to both country-specific and binational efforts to ensure adequate resource security. For example, the passage of the binational 1944 Mexican Water Treaty and the establishment of the International Boundary and Water Commission, the entity charged with determining the most effective way to execute the treaty by allocating surface water across shared river systems, have been viewed as successes in efforts to mitigate potential water disputes. However, although the 1944 Water Treaty marked progress in binational surface water management, it largely ignored the protection of groundwater resources and water quality, which continue to be large challenges for both countries.

Some existing binational initiatives have improved the planning, development, and implementation of cross-border environmental programs and infrastructure, and have resulted in increased access to drinking water, more effective management of wastewater flows, improved air quality, and better solid waste management. Despite examples of progress, there are still many areas of binational partnership that could facilitate better water resource management and offer a large potential for coordination.

Building desalination facilities to treat flows of wastewater and generate more potable water supplies is another area for expanding binational collaboration. Attempts to improve progress in these areas have faced well-known hurdles. Differences in regulation between countries, as well as increased population, urbanization, and industrialization, have complicated some efforts to manage shared water resources, particularly under the increasing challenges posed by climate change.

Other areas of the economy hold potential for binational collaboration. The coordination between such industries as energy and mining is still nascent, but both industries have placed increasing priority on environmental sustainability and social responsibility, presenting large opportunities for progress. The border region is particularly rich in energy resources, both in terms of renewable and nonrenewable sources. The vast renewable energy potential, in particular, could provide opportunities for binational grid expansion, which could facilitate larger penetrations of intermittent wind and solar generation resources to be integrated into a binational grid so that electricity could be traded more easily across borders. By diversifying and expanding the regional extent of the power grid, the challenges posed by the intermittencies of these variable renewable energy generators could be mitigated, since a large regional grid would be less vulnerable to local lapses in wind or solar resource availability. Similarly, while the United States has greatly expanded the use of hydraulic fracturing techniques for oil and gas development, particularly in Texas, the production of shale reservoirs in Mexico is nascent. Energy reforms in Mexico over time have vastly increased the potential for binational cooperation in the energy space to spur synergistic benefits for both countries.

FINDINGS AND CONCLUSIONS

FINDING 1: Although the U.S.–Mexico region includes a diversity of habitats, its mostly arid landscape and the depletion of surface and groundwater supplies are the cause of significant binational water stress.

FINDING 2: Since at least the 1990s, U.S.–Mexico binational partnerships for environmental conservation have improved the planning, development, and implementation of cross-border environmental programs and infrastructure.

FINDING 3: The U.S.–Mexico border dissects the lands of approximately 60 Indigenous nations. Though the border has split the communities in two, many of the nations still maintain close cross-border cultural, economic, and political ties.

FINDING 4: The border region has seen a significant population increase in recent years due to heightened migration and industrialization. Sprawling urbanization has led to the development of formal and informal communities that often lack adequate infrastructure, place significant stress on natural resources, and jeopardize the public health and security of residents.

FINDING 5: Though they share environmental conditions, the "mirror cities" along the U.S.–Mexico border differ widely in their infrastructure, resource management, economic status, legislation, and culture. Historically, these differences complicate the binational management of shared resources, such as groundwater and wastewater treatment.

FINDING 6: The U.S.–Mexico border region is at the center of a binational network of renewable and nonrenewable energy flow. Mexico has historically relied heavily on the United States for nonrenewable energy, but the growth of renewable energy in Mexico due to reforms in the past decade has changed the energy landscape on both sides of the border.

FINDING 7: Water is one of the most consequential resources in the binational region. The 1944 Water Treaty guided surface water management, but groundwater management and water quality continue to be issues in both countries. Differences in water regulation between countries, as well as increased population, urbanization, and industrialization, also complicate shared water management.

CONCLUSION 1: The U.S.–Mexico border region faces many ongoing challenges in safeguarding the sustainability of its natural resources—scarce in some aspects yet abundant in others—to ensure the economic vitality and livelihoods of its people while protecting its cultural richness and unique natural environment.

The binational region is experiencing increasing interactions of people and commerce, the growing interdependence of the two countries on water stocks and flows, and expanding ecological linkages. The region's sustainability challenges are exacerbated by stressors, such as global climate change, increasing urbanization and industrialization, and population and economic growth.

CONCLUSION 2: There is growing potential for partnership efforts around binational industrial, energy, and mining sustainability.

The movement of maquiladoras/assembly plants toward renewable energy resources is altering the industrial landscape of the region. While a relatively recent phenomenon, the maquiladora/assembly plants-dominated industrial environment at the border is being altered by the moves toward renewable energies and the ever-changing political imperatives in both countries. The increased focus on environmental sustainability and social responsibility, as well as the shift in mining technology and innovation, allow for partnerships with multiple actors and across sectors to address binational challenges.

CONCLUSION 3: Navigating the sustainability challenges in the U.S.–Mexico border region will require sound governance and the building and strengthening of strategic partnerships.

Strategic partnerships, engaging a diversity of stakeholders on either side of the border, are needed to devise strategies that both support the region's sustainable development and protect the well-being of humans and ecosystems within it.

REFERENCES

Aguilar-Pesantes, A. Peña Carpio, E. Vitvar, T. Koepke, R., and Menéndez-Aguado, J.M. (2021). A comparative study of mining control in Latin America. *Mining*, 1(1), 6–18. doi: 10.3390/mining1010002.

Albrecht, T.R., Varady, R.G., Zuniga-Teran, A.A., Gerlak, A.K., Routson De Grenade, R., Lutz-Ley, A., Martín, F., Megdal, S.B., Meza, F., Ocampo Melgar, D., Pineda, N., Rojas, F., Taboada, R., and Willems, B. (2018). Unraveling transboundary water security in the arid Americas. *Water International*, 43(8), 1075–1113. doi: 10.1080/02508060.2018.1541583.

Armienta, M.A., and Segovia, N. (2008). Arsenic and fluoride in the groundwater in Mexico. *Environmental Geochemistry and Health*, 30, 345–353.

Barker, R., Scott, C.A., de Fraiture, C., and Amarasinghe, U. (2000). Global water shortages and the challenge facing Mexico. *International Journal of Water Resources Development*, 16(4), 525–542. doi: 10.1080/713672542.

Batalova, J., Blizzard, B., and Bolter, J. (2020). *Frequently Requested Statistics on Immigrants and Immigration in the United States.* Migration Policy Institute. February 14. Available: https://www.migrationpolicy.org/article/frequently-requested-statistics-immigrants-and-immigration-united-states.

Beaver, J.C. (2007). *U.S. International Borders: Brief Facts.* CRS Report for Congress. RS21729. February 1. Washington, DC: Congressional Research Service.

Beittel, J.S. (2020). *Mexico: Organized Crime and Drug Trafficking Organizations.* CRS Report R41576. July 28. Washington, DC: Congressional Research Service.

Bohn, T.J., Vivoni, E.R., Mascaro, G., and White, D.D. (2018). Land and water use changes in the US-Mexico border region, 1992–2011. *Environmental Research Letters*, 13(11), 1–8. Available: https://doi.org/10.1088/1748-9326/aae53e.

Bonner, R., and Rozental, A. (2009). *Managing the United States-Mexico Border: Cooperative Solutions to Common Challenges: Full Report of the Binational Task Force on the United States-Mexico Border.* Available: https://www.wilsoncenter.org/publication/managing-the-united-states-mexico-border-cooperative-solutions-to-common-challenges.

Buono, R.M., and Eckstein, G. (2014). Minute 319: A cooperative approach to Mexico–US hydro-relations on the Colorado River. *Water International, 39*(3), 263–276.

CalEPA (California Environmental Protection Agency). (2020). *New River Strategic Plan Updates*. Available: https://calepa.ca.gov/border-affairs-program/new-river-strategic-plan-updates/.

Callegary, J.B., Minjárez Sosa, I., Tapia Villaseñor, E.M., dos Santos, P., Monreal Saavedra, R., Grijalva Noriega, F.J., Huth, A.K., Gray, F., Scott, C.A., Megdal, S.B., Oroz Ramos, L.A., Rangel Medina, M., and Leenhouts, J.M. (2016). *Binational Study of the Transboundary San Pedro Aquifer.* United States and Mexico: International Boundary and Water Commission.

Campbell, H. (2007). El narco-folklore: Narrativas e historias de la droga en la frontera. *Nóesis. Revista de Ciencias Sociales y Humanidades, 16*(32), 46–70. Available: https://www.redalyc.org/pdf/859/85903203.pdf.

Carrillo, G., Uribe, F., Lucio, R., Ramirez Lopez, A., and Korc, M. (2017). The United States–Mexico border environmental public health: The challenges of working with two systems. *Revista Panamericana Salud Publica, 41*, 1–7.

Carter, N.T., Mulligan, S.P., and Seelke, C.R. (2017). *U.S.-Mexico Water Sharing: Background and Recent Developments.* Washington, DC: Congressional Research Service.

CEC (Commission for Environmental Cooperation). (2004). *North American Air Quality and Climate Change Standards, Regulations, Planning and Enforcement at the National, State/Provincial and Local Levels*. Available: http://www3.cec.org/islandora/fr/item/2145-north-american-air-quality-and-climate-change-standards-regulations-planning-and-en.pdf.

_____. (2011). *North American Terrestrial Ecoregions–Level III*. Available: http://www3.cec.org/islandora/en/item/10415-north-american-terrestrial-ecoregionslevel-iii.

Córdova, A., and de la Parra, C.A. (2007). *Una barrera a nuestro ambiente compartido: El muro fronterizo entre México y Estados Unidos*. El Colegio de la Frontera Norte: Secretaría de Medio Ambiente y Recursos Naturales. Instituto Nacional de Ecología: Consorcio de Investigación y Política Ambiental del Suroeste.

Cresswell, A., Burke, G.B., and Navarrete, C. (2009). *Mitigating Cross-Border Air Pollution: The Power of a Network*. Albany, NY: Center for Technology in Government, University of Albany, SUNY. Available: www.ctg.albany.edu/publications/reports/jac_mitigating.

CRS (Congressional Research Service). (2017). U.S.-Mexico water sharing: Background and recent developments.

_____. (2020). The North American Development Bank. *In Focus*, June 18. Available: https://crsreports.congress.gov/product/pdf/IF/IF10480.

Díaz, E. (2009). Mercado de trabajo e industria maquiladora en Sonora y la frontera norte. *Región y Sociedad, 21*(44), 43–70.

Díaz-Caravantes, R.E., and Wilder, M. (2014). Water, cities and peri-urban communities: Geographies of power in the context of drought in Northwest Mexico. *Water Alternatives, 7*(3), 499–517.

Eades, L. (2018). *Air Pollution at the U.S.–Mexico Border: Strengthening the Framework for Bilateral Cooperation.* Princeton, NJ: Association of Professional Schools and International Affairs and The Woodrow Wilson School of Public and International Affairs, Princeton University.

Eckstein, G.E. (2011). Buried treasure or buried hope? The status of Mexico-U.S. transboundary aquifers under international law. *International Community Law Review, 13*, 273–290. doi: 10.1163/187197311X582395.

EIA (U.S. Energy Information Administration). (2020a). *International Energy Statistics*. Available: https://www.eia.gov/international/overview/world.

_____. (2020b). *Monthly Energy Review*, Tables 1.3 and 10.1. April 2020, preliminary data. Available: https://www.eia.gov/energyexplained/us-energy-facts/.

_____. (2020c). *U.S. States State Profiles and Energy Estimates.* Available: https://www.eia.gov/state/.

_____. (2020d). *Country Analysis. Executive Summary: Mexico.* Available: https://www.eia.gov/international/analysis/country/MEX.

_____. (2020e). *In 2019, the U.S. Imported $13 Billion of Energy Goods from Mexico, Exported $34 Billion.* Available: https://www.eia.gov/todayinenergy/detail.php?id=45756.

El Colef. (2019). La era de Trump y sus impactos en la frontera norte de México. Economía, Población y Desarrollo. *Cuadernos de Trabajo,* 49(2), 1–35. Universidad Autónoma de Ciudad Juárez.

EPA (U.S. Environmental Protection Agency). (1996). *US/Mexico Border XXI Program: Framework Document.* Available: https://catalog.hathitrust.org/Record/101234357.

EPA-SEDUE (U.S. Environmental Protection Agency and Secretaría de Desarrollo Urbano y Ecología). (1991). *Integrated Environmental Plan for the Mexico-U.S. Border Area (First Stage, 1992–1994): Working Draft.* Available: https://nepis.epa.gov/Exe/ZyPURL.cgi?Dockey=91017YO0.txt.

EPA-SEMARNAT (U.S. Environmenal Protection Agency and Secretaría de Medio Ambiente y Recursos Naturales). (2012). *Border 2020 U.S-Mexico Environmental Program (Summary).* Available: https://www.epa.gov/sites/production/files/documents/border2020summary.pdf.

GAO (U.S. Government Accountability Office). (2018). *North American Energy Integration: Information about Cooperation with Canada and Mexico and Among U.S. Agencies.* Report to the Subcommittee on the Western Hemisphere, Committee on Foreign Affairs, House of Representatives. GAO-18-575. Washington, DC.

Gerlak, A.K. (2006). Federalism and the U.S. water policy: Lessons for the twenty-first century. *Publis,* 26(2), 231–257.

Giner, M.E., Córdova, A., Vázquez-Gálvez, F.A., and Marruffo, J. (2019). Promoting green infrastructure in Mexico's northern border: The Border Environment Cooperation Commission's experience and lessons learned. *Journal of Environmental Management,* 248(10), 109–104. doi: 10.1016/j.jenvman.2019.06.005.

GNEB (Good Neighbor Environmental Board). (2014). *Ecological Restoration in the U.S.-Mexico Border Region.* 16th Report of the Good Neighbor Environmental Board to the President and Congress of the United States. Publication number EPA 130-R-14-001. Available: https://www.epa.gov/sites/production/files/2016-12/documents/16th_gneb_report_english_final_web.pdf.

Goetz, R.U., and Berga, D. (Eds.). (2006). *Frontiers in Water Resource Economics* (Vol. 29). Springer Science & Business Media. New York, NY: Springer Science & Business Media.

Greenwald, N., Segee, B., Curry, T. and Bradley, C. (2017). *A Wall in the Wild: The Disastrous Impacts of Trump's Border Wall on Wildlife.* Tucson, AZ: Center for Biological Diversity.

Gross, S. (2019). Order from Chaos: AMLO reverses positive trends in Mexico's energy industry. *Brookings* blog, December 20. Available: https://www.brookings.edu/blog/order-from-chaos/2019/12/20/amlo-reverses-positive-trends-in-mexicos-energy-industry/.

Hernández Pérez, J.J. (2019). Sistema de innovación agrícola como estrategia de competitividad de los productores sonorenses en el contexto del TLCAN. *Estudios Sociales,* 29(54), 2–35. doi: 10.24836/es.v29i54.828.

HHS (U.S. Department of Health and Human Services). (2017). *The U.S.-Mexico Border Region.* Available: https://www.hhs.gov/about/agencies/oga/about-oga/what-we-do/international-relations-division/americas/border-health-commission/us-mexico-border-region.

HUD (U.S. Department of Housing and Urban Development). (2020). *Colonias History.* Available: https://www.hudexchange.info/programs/cdbg-colonias/colonias-history/.

Huesca Reynoso, L., and Llamas Rembao, L.I. (2019). Crisis y resiliencia en género y salarios: el sector manufacturero en México y la frontera norte. *Frontera Norte, 31*(16), 1–23. doi: 10.33679/rfn.v1i1.2051.

Hruska, T. (2020). Evolving patterns of agricultural frontier expansion in Mexico's Chihuahuan Desert: A political ecology approach. *Journal of Land Use Science, 15*(2-3), 270–289. doi: 10.1080/1747423x.2019.1646332.

IEA (International Energy Agency). (2020a). *Total Energy Supply (TPES) by Source, United States, 1990–2019.* Available: https://www.iea.org/countries/united-states.

———. (2020b). *Total Energy Supply (TPES) by Source, Mexico, 1990–2019.* Available: https://www.iea.org/countries/mexico.

Israel, E., and Batalova, J. (2020). *Mexican Immigrants in the United States.* Available: https://www.migrationpolicy.org/article/mexican-immigrants-united-states-2019.

Jackson, R.B., Vengosh, A., Darrah, T.H., Warner, N.R., Down, A., Poreda, R.J., Osborn, S.G., Zhao, K., and Karr, J.D. (2013). Increased stray gas abundance in a subset of drinking water wells near Marcellus shale gas extraction. *Proceedings of the National Academy of Sciences, 110*(28), 11250–11255. doi: 10.1073/pnas.1221635110.

James, I. (2019). Poisoned cities deadly border: This city's air is killing people. Who will stop it? *Desert Sun,* January 15. Available: https://www.desertsun.com/in-depth/news/environment/border-pollution/poisoned-cities/2018/12/05/air-pollution-taking-deadly-toll-u-s-mexico-border/1381585002/.

Jepson, W. (2014). Measuring 'no-win' waterscapes: Experience-based scales and classification approaches to assess household water security in *colonias* on the U.S.-Mexico border. *Geoforum, 51,* 107–120. doi: 10.1016/j.geoforum.2013.10.002.

King, C.W., Stillwell, A.S., Twomey, K.M., and Webber, M.E. (2013). Coherence between water and energy policies. *Natural Resources Journal, 53*(1), 117–215.

Maganda, C. (2005). Collateral damage: How the San Diego-Imperial Valley water agreement affects the Mexican side of the border. *The Journal of Environment and Development, 14,* 486–506. doi: 10.1177/1070496505282668.

McCallum, J.W., Rowcliffe, J.M., and Cuthill, I.C. (2014). Conservation on international boundaries: The impact of security barriers on selected terrestrial mammals in four protected areas in Arizona, USA. *PLoS ONE, 9*(4), e93679. doi: 10.1371/journal.pone.0093679.

Megdal, S.B., and Scott, C.A. (2011). The importance of institutional asymmetries to the development of binational aquifer assessment programs: The Arizona-Sonora experience. *Water, 3,* 949–963. doi: 10.3390/w3030949.

Mendoza-Lagunas, J.L., Meza-Figueroa, D.M., Martínez-Cinco, M.A., O'Rourke, M.K., Centeno-García, E., Romero, F.M., García-Rico, L., and Meza-Montenegro, M.M. (2019). Health risk assessment in children by arsenic and mercury pollution of groundwater in a mining area in Sonora, Mexico. *Journal of Geoscience and Environment Protection, 7*(6), 90–105. doi: 10.4236/gep.2019.76008.

Miller, B., Schaetzl, R., and Frank, J. (2012). *The Soil Productivity Index: Taxonomically Based, Ordinal Estimates of Soil Productivity.* New York: Association of American Geographers Annual Meeting. doi: 10.13140/2.1.3036.5127.

Mohtar, R.H., Shafiezadeh, H., Blake, J., and Daher, B. (2019). Economic, social, and environmental evaluation of energy development in the Eagle Ford shale play. *Science of the Total Environment, 646*(1), 1601–1614. Available: https://doi.org/10.1016/j.scitotenv.2018.07.202.

Mumme, S.P. (2005). Advancing binational cooperation in transboundary aquifer management on the U.S.-Mexico Border. *Colorado Journal of International Environmental Law and Policy, 16*(1), 77–94.

———. (2020). The 1944 Water Treaty and the incorporation of environmental values in US-Mexico transboundary water governance. *Environmental Science and Policy, 112,* 126–133.

NASEM (National Academies of Sciences, Engineering, and Medicine). (2018). *Advancing Sustainability of U.S.-Mexico Transboundary Drylands: Proceedings of a Workshop*. Washington, DC: The National Academies Press. doi: 10.17226/25253.

Osuchukwu, O., Nuñez, M., Packard, S., Ehiri, J., Rosales, C., Hawkins, E., Avilés, J.G.G., and Oren, E. (2017). Latent tuberculosis infection screening acceptability among migrant farmworkers. *International Migration*, 55(5), 62–74.

Pavlakovich-Kochi, V. (2006). The Arizona-Sonora Region: A decade of transborder region building. *Estudios Sociales*, 14(27), 25–55. Available: http://www.scielo.org.mx/scielo.php?script=sci_arttext&pid=S0188-45572006000100002&lng=es&tlng=en.

Peer, R.A.M., and Sanders, K.T. (2018). The water consequences of a transitioning US power sector. *Applied Energy*, 210(15), 613–622. doi: 10.1016/j.apenergy.2017.08.021.

Pelozzi, K., Kozo, J., Ferran, K., Wooten, W., Rangel Gomez, G., and AL-DeLaimy, W.K. (2014). One Bioregion/One Health: An integrative narrative for transboundary planning along the US–Mexico border. *Global Society*, 28(4), 419–440. doi:10.1080/13600826.2014.951316.

Peña Muñoz, J.J. (2018). Recomposición de la migración laboral en la frontera norte de México. *Frontera Norte*, 30(59), 81–102. doi: 10.17428/rfn.v30i59.645.

Pérez Ortega, R. (2020). Pools in the Mexican desert are a window into Earth's early life. *Science*, June 30. Available: https://www.sciencemag.org/news/2020/06/pools-mexican-desert-are-window-earth-s-early-life.

Peters, R., Ripple, W.J., Wolf, C., Moskwik, M.,Carreón-Arroyo, G., Ceballos, G., Córdova, A., Dirzo, R., Ehrlich, P.R., Flesch, A.D., List, R., Lovejoy, T.E., Noss, R.F., Pacheco, J., Sarukhán, J.K.,Soulé, M.E., Wilson, E.O., and Miller, J.R.B. (2018). Nature divided, scientists united: US-Mexico border wall threatens biodiversity and binational conservation. *BioScience*, 68(10), 740–743. Available: https://doi.org/10.1093/biosci/biy063.

Peters, R.L., and Clark, M. (2018). *In The Shadow of the Wall: Park II—Borderlands Conservation Hotspots on the Line*. Washington, DC: Defenders of Wildlife.

Piña Osuna, F.M., and Poom Medina, J. (2019). Deterioro social y participación en el tráfico de drogas en el estado de Sonora. *Frontera Norte* 31(1), 1–20. doi: 10.33679/rfn.v1i1.1976.

Rahm, B.G., and Riha, S.J. (2014). Evolving shale gas management: Water resource risks, impacts, and lessons learned. *Environmental Science: Processes and Impacts*, 16, 1400–1412.

Ramos, J.M., and Reyes, M. (2006). Organizaciones no gubernamentales y la contaminación del aire en la frontera de Baja California, Mexico-California, Estados Unidos. Contexto y desafíos. *Región y Sociedad*, 18(37), 37–84.

Registro Agrario Nacional. (2019). *Datos Geográficos Perimetrales de los Núcleos Agrarios Certificados, por Estado*. Available: http://datos.gob.mx.

Rodríguez, J.L.S. (2016). Matrices indígenas del norte de México. *Desacatos. Revista De Ciencias Sociales*, 50, 172–183. doi: 10.29340/50.1548.

Sanchez, R., and Eckstein, G. (2020). Groundwater management in the borderlands of Mexico and Texas: The beauty of the unknown, the negligence of the present, and the way forward. *Water Resources Research*, 56(3), e2019WR026068. doi: 10.1029/2019WR026068.

Sanchez, R., Lopez, V., and Eckstein, G. (2016). Identifying and characterizing transboundary aquifers along the Mexico-US border: An initial assessment. *Journal of Hydrology*, 535, 101–119. doi: 10.1016/j.jhydrol.2016.01.070.

Sanders, K.T., King, C.W., Stillwell, A.S., and Webber, M.E. (2013). Clean energy and water: Assessment of Mexico for improved water services and renewable energy. *Environment, Development and Sustainability*, 15, 1303–1321. doi: 10.1007/s10668-013-9441-5.

Sandoval-Solis, S., Teasley, R.L., McKinney, D.C., Thomas, G.A., and Patiño-Gomez, C. (2013). Collaborative modeling to evaluate water management scenarios in the Rio Grande Basin. *Journal of the American Water Resources Association*, 49(3), 639–653.

Santamaría, A. (2012). *Las Jefas del Narco*. México: Grijalbo.

SCERP (Southwest Center for Environmental Research and Policy). (2004). Indigenous groups of Mexico's Northern Border Region. In M. Wilken-Robertson (Ed.), *The U.S.-Mexican Border Environment: Tribal Environmental Issues of the Border Region.* SERP Monograph series No 9. San Diego, CA: San Diego State University Press.

Schur, E.L. (2017). Potable or affordable? A comparative study of household water security within a transboundary aquifer along the U.S.-Mexico border. *Journal of Latin American Geography,* 16(3), 29–58. doi: 10.1353/lag.2017.0051.

Scott, C.A. 2013. Electricity for groundwater use: Constraints and opportunities for adaptive response to climate change. *Environmental Research Letters,* 8. doi: 10.1088/1748-9326/8/3/035005.

Scott, C.A., Vicuña, S., Blanco-Gutiérrez, I., Meza, F., and Varela-Ortega, C. (2014). Irrigation efficiency and water-policy implications for river-basin resilience. *Hydrology and Earth System Sciences,* 18, 1339–1348, doi: 10.5194/hess-18-1339-2014.

Scott, C.A., Megdal, S., Oroz, L.A., Callegary, J., and Vandervoet, P. (2012). Effects of climate change and population growth on the transboundary Santa Cruz aquifer. *Climate Research,* 51, 159–170. doi: 10.3354/cr01061.

Scott, C.A., and Banister, J.M. (2008). The dilemma of water management 'regionalization' in Mexico under centralized resource allocation. *Water Resources Development,* 24(1), 61–74.

Secretaría de Medio Ambiente y Recursos Naturales and U.S. Enviromental Protection Agency (2012). *Border 2020: U.S.-Mexico Environmental Program.* Available: www.epa.gov/Border2020.

Secretaría de Turismo. (2019). *Compendio Estadístico del Turismo en México 2019.* Subsecretaría de Planeación–Dirección General de Información y Análisis. Available: https://www.datatur.sectur.gob.mx/SitePages/CompendioEstadistico.aspx.

Shah, T., Scott, C., and Buechler, S. (2004). Water sector reforms in Mexico: Lessons for India's new water policy. *Economic and Political Weekly,* 39(4), 361–370. Available: http://www.jstor.org/stable/4414554.

Shaji, E., Santosh, M., Sarath, K.V., Prakash, P., Deepchand, V., and Divya, B.V. (2020). Arsenic contamination of groundwater: A global synopsis with focus on the Indian Peninsula. *Geoscience Frontiers.* doi: 10.1016/j.gsf.2020.08.015.

SIAP (Sistema de Información Agroalimentaria y Pesquera). (2019). *Cierre de la Producción Pecuaria (1980–2019).* Available: http://nube.siap.gob.mx/cierre_pecuario/.

_____. (2018a). *Anuario Estadístico de la Producción Agrícola.* Available: https://nube.siap.gob.mx/cierreagricola/.

_____. (2018b). *Estadística de la Producción Agrícola de 2018.* Available: http://infosiap.siap.gob.mx/gobmx/datosAbiertos.php.

Sietz, D., Lüdeke, M.K.B., and Walther, C. (2011). Categorisation of typical vulnerability patterns in global drylands. *Global Environmental Change,* 21(2), 431–440. doi: 10.1016/j.gloenvcha.2010.11.005.

Soden, D.L. (2006). *At the Cross Roads: US / Mexico Border Counties in Transition.* IPED Technical Reports, Paper 27. El Paso: University of Texas at El Paso, Institute for Policy and Economic Development (IPED). Available: http://digitalcommons.utep.edu/iped_techrep/27.

Solís, M., and Ávalos, M. (2017). Construyendo ciudadanía laboral en la frontera norte de México. *Trabajo y Sociedad,* 29, 287–305. Available: https://www.redalyc.org/articulo.oa?id=387352369015.

Spring, Ú.O. (2016). The water, energy, food and biodiversity nexus: New security issues in the case of Mexico. In H. Brauch, Ú.O. Spring, J. Bennett, and O.S. Serrano (Eds.), *Addressing Global Environmental Challenges from a Peace Ecology Perspective* (4th ed., pp. 113–144). Switzerland: Springer International Publishing. doi: 10.1007/978-3-319-30990-3_6.

Talmage, C.A., Pjawka, D., and Hagen, B. (2019). Re-examination of quality of life indicators in US-Mexico border cities: A critical review. *International Journal of Community Well-Being, 2*, 135–154.

USDA (U.S. Department of Agriculture). (2017). *Census Agriculture Atlas Maps.* Available: https://www.nass.usda.gov/Publications/AgCensus/2017/Online_Resources/Ag_Atlas_Maps/index.php.

U.S. Department of the Interior, Bureau of Reclamation. (2013). *The Colorado River Basin Water Supply and Demand Study Fact Sheet.* Available: https://www.usbr.gov/lc/region/programs/crbstudy/FactSheet_June2013.pdf.

_____. (2016a). Rio Grande Basin: Reclamation managing water in the west. Chapter 7 in *SECURE Water Act Section 9503(c) — Reclamation Climate Change and Water 2016.* Available: https://www.usbr.gov/climate/secure/docs/2016secure/2016SECUREReport.pdf.

_____. (2016b). Colorado River Basin. Reclamation managing water in the west. Chapter 3 in *SECURE Water Act Section 9503(c) — Reclamation Climate Change and Water 2016.* Available: https://www.usbr.gov/climate/secure/docs/2016secure/2016SECUREReport.pdf.

Varady, R.G., and Ward, E. (2009). Transboundary conservation in the borderlands: What drives environmental change? In L. Lopez-Hoffman, E. McGovern, R.G. Varady, and K.W. Flessa (Eds.), *Conservation of Shared Environments: Learning from the United States and Mexico* (pp. 17–25). Tucson, AZ: University of Arizona Press.

Vásquez-León, M., West, C.T., Wolf, B., Moody, J., and Finan, T.J. (2002). Vulnerability to climate variability in the farming sector: A case study of groundwater-dependent agriculture in southeastern Arizona. *CLIMAS Report Series CL1-02.* Tuscon, AZ: Climate Assessment for the Southwest, The University of Arizona.

Vidaurrázaga Obezo, F.R. (2003). Los cambios en la política agropecuaria y la propiedad social rural en la frontera norte. *Estudios Fronterizos, 4*(8), 163–188.

Vidic, R.D., Brantley, S.L., Vandenbossche, J.M., Yoxtheimer, D., and Abad, J.D. (2013). Impact of shale gas development on regional water quality. *Science, 340*(6134). doi: 10.1126/science.1235009.

Vietor, R.H.K., and Sheldahl-Thomason, H. (2017). Mexico's energy reform. *Harvard Business School Case 717-1,207.* Boston, MA: Harvard Business School Publishing.

Villarejo, D. (2002). The health of U.S. hired farmworkers. *Annual Review of Public Health, 24*, 175–93. doi: 10.1146/annurev.publhealth.24.100901.140901.

von Lüpke, H., and Well, W. (2019). Analyzing climate and energy policy integration: The case of the Mexican energy transition. *Climate Policy.* doi: 10.1080/14693062.2019.1648236.

Weijermars, R., Sorek, N., Sen, D., and Ayers, W.B. (2017). Eagle Ford Shale play economics: U.S. versus Mexico. *Journal of Natural Gas Science and Engineering, 38*, 345–372. doi: 10.1016/j.jngse.2016.12.009.

Wiken, E., Jiménez Nava, F., and Griffith, G. (2011). *North American Terrestrial Ecoregions. Level III.* Montreal, Canada: Commission for Environmental Cooperation.

Wilder, M.O., Aguilar-Barajas, I., Pineda-Pablos, N., Varady, R.G., Megdal, S.B., McEvoy, J., Merideth, R., Zúñiga-Terán, A.A., and Scott, C.A. (2016). Desalination and water security in the US–Mexico border region: Assessing the social, environmental and political impacts. *Water International, 41*(5), 756–775. doi: 10.1080/02508060.2016.1166416.

Wilder, M., Garfin, G., Ganster, P., Eakin, H., Romero-Lankao, P., Lara-Valencia, F., Cortez-Lara, A.A., Mumme, S., Neri, C., and Muñoz-Arriola, F. (2013). Climate change and U.S.-Mexico border communities. In G. Garfin, A. Jardine, R. Merideth, M. Black, and S. LeRoy (Eds.), *Assessment of Climate Change in the Southwest United States: A Report Prepared for the National Climate Assessment,* (pp. 340–384). Washington, DC: Island Press. doi: 10.5822/978-1-61091-484-0.

Wilder, M., and Romero, P. (2006). Paradoxes of decentralization: Water reform and social implications in Mexico. *World Development, 34*(11), 1977–1995.

Wilder, M., Scott, C.A., Pablos, N.P., Varady, R.G., Garfin, G.M., and McEvoy, J. (2010). Adapting across boundaries: Climate change, social learning, and resilience in the U.S.-Mexico border region. *Annals of the Association of American Geographers, 100*(4), 917–928. doi: 10.1080/00045608.2010.500235.

Zuniga-Teran, A.A., Mussetta, P.C., Lutz Ley, A.N., Díaz-Caravantes, R.E., and Gerlak, A.K. (2020). Analyzing water policy impacts on vulnerability: Cases across the rural-urban continuum in the arid Americas. *Environmental Development.* doi: 10.1016/j.envdev.2020.100552.

Appendix E

Acronym List

CalEPA	California Environmental Protection Agency
CEC	Commission for Environmental Cooperation
CRS	Congressional Research Service
EIA	U.S. Energy Information Agency
EPA	U.S. Environmental Protection Agency
EPA-SEDUE	U.S. Environmental Protection Agency and Secretaría de Desarrollo Urbano y Ecología
EPA-SEMARNAT	U.S. Environmental Protection Agency and Secretaría del Medio Ambiente y Recursos Naturales
GAO	U.S. Government Accountability Office
GNEB	Good Neighbor Environmental Board
HHS	U.S. Department of Health and Human Services
HUD	U.S. Department of Housing and Urban Development
IEA	International Energy Agency
INEGI	Instituto Nacional de Estadística, Geografía e Informática

NASEM	National Academies of Science, Engineering, and Medicine
SCERP	Southwest Center for Environmental Research and Policy
USDA	U.S. Department of Agriculture